AF874323

MODERN TEX and Its Applications

FHB Lüneburg/W 962
014 147 X

Michael Vulis

FHS Nordostniedersachsen
Bibliothek Fachbereich Wirtschaft
Zug.-Nr. 1187/98
Signatur EDV 110.1
39,1
Geschenk +1 Disk.

CRC

CRC Press
Boca Raton Ann Arbor London Tokyo

Library of Congress Cataloging-in-Publication Data

Vulis, Michael.
Modern TeX and its applications / Michael Vulis.
p. cm.
Includes index.
ISBN 0-8493-4431-X
1. TeX (Computer file). 2. Computerized typesetting — Computer programs. 3. Mathematics printing — Computer programs. I. Title.
Z253.4.T47V84 1993
686.2′2544536—dc20

92-40781
CIP

This book represents information obtained from authentic and highly regarded sources. Reprinted material is quoted with permission, and sources are indicated. A wide variety of references are listed. Every reasonable effort has been made to give reliable data and information, but the author and the publisher cannot assume responsibility for the validity of all materials or for the consequences of their use.

Neither this book nor any part may be reproduced or transmitted in any form or by any means, electronic or mechanical, including photocopying, microfilming, and recording, or by any information storage and retrieval system, without permission in writing from the publisher.

Direct all inquiries to CRC Press, Inc., 2000 Corporate Blvd., N. W., Boca Raton, Florida, 33431.

© 1993 by CRC Press, Inc.

International Standard Book Number 0-8493-4431-X

Library of Congress Card Number 92-40781
Printed in the United States 2 3 4 5 6 7 8 9 0
Printed on acid-free paper

Contents

1. Introduction 1

1.1 About this book and TeX 3

1.2 TeX, VTeX, and PLAIN 4

1.3 Organization of the Book 6

1.4 Basics 7

1.5 More Basics 12

1.6 Introduction to Error Handling 14

1.7 Input Files 16

2. Fonts and Characters 17

2.1 Typesetting: A Historical Overview 19

2.2 Font Representation 21

2.3 Standard Fonts 23

2.4 Groups 26

2.5 Underlining 28

2.6 Other Fonts 30

2.7 Font Sizes and Magnification 34

2.8 Font Attributes 38

2.9 Font Compatibility and Aliasing 43

2.10 Character Set: Letters, Digits, Punctuation 45

2.11 Character Set: Special Characters 46

2.12 Accents and Foreign Characters 49

2.13 Use of Cyrillic Fonts 51

3. Formatting 55

3.1 Page Layout 57

3.2 Lines of Text ... 59

3.3 Headers and Footers. Page Numbers ... 60

3.4 Default Horizontal Spacing ... 63

3.5 Horizontal Skips ... 65

3.6 Paragraph Formatting ... 67

3.7 Non-Standard Paragraph Formatting ... 69

3.8 Lists ... 72

3.9 Line Breaks ... 75

3.10 Hyphenation ... 76

3.11 Overfull and Underfull Lines ... 78

3.12 Vertical Spacing ... 84

3.13 Vertical Skips ... 85

3.14 Page Breaks ... 87

3.15 Footnotes ... 88

3.16 Verbatim Printing ... 90

4. Math Mode ... 93

4.1 Ordinary Math and Display Math ... 95

4.2 Subscripts and Superscripts ... 96

4.3 Mixing Text with Math ... 98

4.4 Sums and Integrals ... 99

4.5 Vertical Stacking. Fractions ... 100

4.6 Standard Function Names ... 104

4.7 Roots ... 105

4.8 Big and Adjustable Delimiters ... 107

4.9 Math Styles ... 110

4.10 Accents, Bars, Hats ... 111

4.11 Lines, Arrows, Captions ... 113

4.12 Matrices ... 116

4.13 Ellipses 119
4.14 Spacing commands 120
4.15 Equation Numbering 121
4.16 Aligning Multi-line Equations 122
4.17 Line Breaks in Math Mode 123
4.18 Symbols. Spacing. Coercion 123
4.19 Phantoms 124
4.20 Math Font Families 126
4.21 More Examples 128

5. Variables, Dimensions, Glue 133

5.1 Variables: General Information 135
5.2 Numbers and Count Variables 137
5.3 Dimensions and Dimensional Variables 141
5.4 Inter-word Spacing. Glue 143
5.5 Infinite Glues. Filling 145
5.6 Vertical Glues 148
5.7 Glue Variables 149

6. Macros 153

6.1 Macros as Abbreviations 155
6.2 Macros with Arguments 157
6.3 Macros with Delimited Arguments 159
6.4 Examples: Macros in Text Mode 162
6.5 Examples: A Simple Stylesheet 163
6.6 Examples: Macros in Math Mode 166
6.7 Scope of a Macro Definition 168
6.8 Controlling Macro Expansion 170
6.9 Conditional Processing: if 171

6.10 Conditional Processing: case 173
6.11 Conditional Processing: User–Defined Tests 175
6.12 Conditional Processing: loops 175
6.13 Category Codes 176
6.14 Changing Category Codes 179
6.15 Using High ASCII Characters 180
6.16 Upper and Lower Cases. Conversion 180
6.17 Accessing Individual Characters 182
6.18 Tokens. Token Strings 182
6.19 Predefined Token Variables 184
6.20 Other Macro Commands 185

7. Input/Output and Extensions 187

7.1 Printing Messages on Screen 189
7.2 File and Terminal Operations 190
7.3 Text References 192
7.4 Indices 193
7.5 Running Other Tools from within TEX 194
7.6 Special Extensions 195

8. Modes, Rules, Boxes 197

8.1 Modes of Operation 199
8.2 Rules 200
8.3 Boxes 202
8.4 Box Variables – Operations on Boxes 206
8.5 Moving Boxes – Kerning 208
8.6 Drop-Case Letters 210
8.7 Shipping out 211
8.8 Filling. Leaders 212

8.9 Gray Rules and Patterns 215

8.10 Broken Underline. Strikeout. Redline 217

9. Tabulation and Tables 219

9.1 Tabulation—Evenly Spaced Tabs 221

9.2 Tabulation—Arbitrary Tabs 223

9.3 Alignment Templates 224

9.4 Using Rules In Tables 228

10. Font Rotation 233

10.1 Circular Typesetting with TEX and METAFONT–by Alan Hoenig 235

10.2 Circular Typesetting with VTEX 240

11. Appendix 245

11.1 Tables of Mathematical Symbols 247

11.2 Text Font Layouts 252

11.3 Line Draw Font 256

11.4 Symbol Font Layouts 257

11.5 MicroPress Vector Fonts—Partial Listing 261

11.6 Defaults 267

11.7 Generating and Using Format Files 268

11.8 Graphics Inclusion in TEX and VTEX 269

11.9 The changes.tex file 271

11.10 Printing Font Charts 278

11.11 Accompanying Software 279

11.12 Comments, Please 281

12. Index 283

Chapter

1

Introduction

1.1 About this book and TEX

Since the introduction of Donald Knuth's TEX in the early 80s the collection of books written about it keeps growing. This collection now includes very detailed and advanced references, books of examples, and an introduction to the language. Some of the books about TEX appeal to computer and TEX professionals (TEXhackers and TEXperts, as prevailing jargon goes), some are aimed at the novices. This book from its inception was designed to cover the (uncovered) middle ground – it's a book for those who want to start using TEX, use TEX, and continue using TEX. It starts with a step-by-step introduction for a beginner; an average user will find it covers and explains most of the questions he may have; while a TEX professional might find this book slow and, possibly, boringly slow-starting, there would be something new even to him.

Though this book's cover shows a single author, many people actually contributed to it. Their contribution came as better phrases, better examples, and better ideas. These were different people with different backgrounds and different thoughts – what united them was that they liked TEX.

Not only did they like TEX; they believed TEX to be a revelation; and they did not hesitate to place the TEXbook on the shelf next to *The Bible*, *The Declaration of Independence*, *Das Kapital* – these were different people and it was TEX that united them.

TEX was a revelation; and as the zealots of the new religion they felt the need to strive to spread the gospel – and this is why this book was written.

They also were concerned the well-being of TEX – and being heretics of a kind, not blinded by the TEX's shining aura – they went on to develop and enhance the system, bringing it more in line with the today's typesetting needs – for TEX was getting old.

Designed more than ten years ago, TEX is a leading contender for the title of the longest surviving software system of any kind. Nevertheless, its age is beginning to show. Static systems are doomed to stagnate and die. Both the strength and the curse of TEX are its stability and uniformity – this is what made it so widespread in the first place – and this is what makes it slowly retreat and retrench as newer technologies come along.

Since its birth, TEX was subjected to only two major heresies – this is in itself a proof of how good the original design was. The first was the Multilingual TEX by Michael Ferguson – an extension of TEX for other-than-English languages. Some of Ferguson's ideas eventually were incorporated into TEX 3.0, adding more life to TEX. The second heresy was the support of modern scalable font technology in VTEX, the MicroPress' implementation of TEX. VTEX enhances TEX in several ways; being the leading deviant dialect, it receives very substantial attention in this volume, since if TEX is going anywhere, it is all but certain that this is the direction it will take.

The following individuals contributed to this edition of this volume

- John Casti, New Mexico Institute, who conceived the idea of this book and followed with it through to the end
- Alan Hoenig, John Jay College of Criminal Justice, who figured out how to rotate fonts
- Sergey Genkin, MicroPress, Inc, and Louis Weinberg, City College and Graduate Center, who proof-read the complete manuscript

Many valuable suggestions and examples are due to

- Jacob Brickman, Texaco
- Derek Glickstein, Wise Choice Software
- Sharon Persinger, Arkansas State University
- Lin Tay, MicroPress, Inc
- Dwen-Ren Tsai, Stockton State College and MicroPress, Intl.

The author is deeply grateful to all of them. On their and his own behalf, it is the author's hope that this book will yet win TEXconverts.

Finally, the author would like to dedicate this book to Lin and Daryl whose `\tolerance` (§3.11) through the process of writing this book made it possible.

1.2 TEX, VTEX, and PLAIN

TEX is a powerful typesetting language developed at Stanford University in the early eighties. The primary designer of TEX, Prof. Donald Knuth, is well known for his many contributions to computer science. TEX was originally developed as a mainframe package and it has been implemented on a large number of platforms, including the IBM 370, DEC VAX, SUN, and APOLLO. These implementations made TEX a *de facto* standard in scientific publishing, since it is more powerful and more flexible than proprietary formatting packages like SCRIPT (IBM) or NROFF (Unix). TEX is also the only typesetting language ever implemented on a variety of different computers. It is not tied to a single operating system and the same syntax and rules are used in almost all of its implementations.

Among other things, TEX can format paragraphs and pages, use a variety of fonts, typeset equations and tables, do automatic hyphenation, and embed pictures and graphs into the text. Many common documents, such as business letters and resumés, are quite easy to typeset and require only a small number of commands, but there are no restrictions (other than physical ones, such as disk space) on the size and complexity of the jobs that TeX can handle. TeX can do lots of different things — and this book will explain most of them.

Besides other praiseworthy things, TeX is a masterpiece of platform-independent and device-independent programming. The philosophy of TeX is that a document should stay the same even if it is moved from a Cray to a PC and even if it is printed today on a dot-matrix printer and tomorrow on a Lino. Toward this device independence, TeX was developed in parallel with a special typeface family — the Computer Modern Fonts.

Thorough system as though it is, TeX is a system developed more than ten years ago. Several features that have become standard in modern word-processors were not included in TeX. These include scalable font support, indexing and referencing, and automatic loading and rescaling of popular graphics formats. These features were added in the VTeX version of the language.

Since this book talks about different dialects all at once, an effort was made to separate the VTeX extensions from the standard Plain language. First, the new VTeX-specific features are always marked with the filled-in check as shown below:

✔ Reverse print is VTeX-specific.

These features are not available under standard implementations of TeX; the reader who uses different brands of TeX should just view them as a kind of a preview of coming attractions. Some other "extensions" described in this book, are not really extensions, but just new macros which any user of TeX can easily add to his implementation. The author considered these macros handy to a point of deserving to be included; he therefore treated them as if they were standard TeX commands and marked them as shown below:

✓ The `\page` command is not standard; use CHANGES.

The macro source for this command is listed in §11.9, and the user who would overcome the challenge of typing in these lines can get them working for himself. VTeX users need not worry about them, since these commands are automatically available in their implementation; they however should be aware that use of these commands may make their documents VTeX-specific.

One more remark that should be made before we plunge into the details. It is that there are actually a few TeX's: Plain TeX, LaTeX™, AMS-TeX™, and some more. These are different macro packages, all based on the same TeX program. The Plain style is the original macro package designed for TeX by Knuth; AMS-TeX enhances Plain for mathematical typesetting; LaTeX is an independent and much larger macro package. Of these three variants this book talks only about the Plain format; the others are excellently described in books by M. Spivak and L. Lamport. To LaTeX users this book is totally inapplicable; to $\mathcal{A}\mathcal{M}\mathcal{S}$-TEX users it falls short of giving the full picture.

Finally, while TEX itself is a very machine-independent, several of the examples in the body of the book demonstrate the interface between the program and the operating system, or refer to disk files. In all such cases the examples are given for the MS-DOS operating system; users of the other systems should make appropriate changes.

1.3 Organization of the Book

This book is divided into chapters, each chapter being subdivided into sections. The first three chapters contain essential information necessary for the proper use of the TEX package; they should be read in order. The remaining chapters describe advanced techniques that can be used in more sophisticated printing jobs. They may be referred to as needed.

TEX has hundreds of commands, most of which are described in this book, and, of course, listed in the Index. Each section lists the commands it covers, as well as other main points, if any, under the heading **TOPICS** ▷, e.g.,

TOPICS ▷ `\centerline`, `\indent`, paragraph formatting.

Important points are occasionally preceded by **NOTE** ▷.

Commands are described in detail, with examples and caveats given. In addition, the syntax of most commands or groups of related commands is symbolically shown under the heading **SYNTAX** ▷,i.e.,

SYNTAX ▷ `\hsize` [=] ⟨number⟩

In a syntax description, optional terms are placed between square brackets '[' and ']', while required material is placed between angle brackets '⟨' and '⟩'. Neither the square brackets nor the angle brackets should be typed as a part of the command.

This book uses the following typographic conventions: file names, TEX messages and the user's input appear in `typewriter font`; the output appears in boxes. The box is not part of the output, it is only used for emphasis. When it is necessary to show that a space is to be inserted, the symbol '␣' will be used instead of a blank space. Whenever a command or other information to be entered is contained between quotes, it must be typed without the quotes unless otherwise directed.

1.4 Basics

TOPICS ▷ default file names, `\bye`, `\end`, `\relax`; running TEX

During its operation TEX reads and writes a number of files, whose specific functions are identified by their default names and/or extensions.

TEX takes its input from one or more ASCII files, which can be created by a text editor or by a word processor capable of writing ASCII files. These files will sometimes be referred to as *source files* or *input files* or simply as *the document*. If you specify just a filename without an extension, TEX assumes the extension to be `.tex`. For example, if you call your source file `article`, TEX will look for a file called `article.tex`. You may, if you wish, choose another extension by specifying a full file name, such as `article.doc`. The source file contains the text to be typeset and TEX *commands* or *control sequences*, which are embedded within the text.

After processing the source file, TEX will write the output, if any, in a file with the extension `.dvi` (which stands for "device-independent"). In the example above, this file will be called `article.dvi`, regardless of the extension of the source file. It contains typesetting information needed by your screen or printer driver.

Along with the output file, TEX also writes a *transcript* or *log file*, whose extension is `.log`. In this example the log file is called `article.log`. It contains detailed information on the job you ran, as well as a list of error messages. The transcript can be useful when correcting errors. This can be facilitated by using a text editor capable of displaying separate windows on the screen, so that the source file and the log file may be viewed simultaneously.

In addition to the three types of files mentioned above, TEX sometimes uses stylesheets. These files have a `.sty` extension (`article.sty`, for example) and contain parameters and commands necessary for a particular kind of job (see §6.5).

TEX can be run in two ways:

- *interactively*—you are prompted for each line of input
- *noninteractively* or *batch*—the input is read from a file you specify when invoking the package (see below).

It is possible to switch between the two modes within the same session.

For interactive use, at the operating system prompt (MS DOS `C>` or `A>`, for example, depending on which disk drive is current), enter

```
C> tex
```

The convention here and subsequently is to show the Operating System prompt, but, of course, it is not typed by the user.

After the logo and other information, TeX will display two asterisks '`**`'. You now have two options:

- you can enter the name of a source file, in which case TeX will proceed automatically (just as in noninteractive mode), or
- you can type '`\relax`' and you will be prompted with one asterisk '`*`'. You are now in interactive mode, where you may enter text or commands line by line. To finish, type

  ```
  \bye
  ```

at the `*`-prompt. TeX will process your input and, if there are no errors, it will display some information about what it did and then return you to the DOS prompt. The output file will be named `texput.dvi` and the transcript file will be `texput.log`.

You can enter any command that starts with the backslash at the `**`-prompt; after this command is processed you will get the `*`-prompt.

Some operating systems, including MS DOS, reserve the backslash character to denote subdirectories. Most versions of TeX, VTeX included, allow you to enter the subdirectories through the slash character. Thus, type

```
**  /mydocs/article
```

instead of

```
**  \mydocs\article
```

With a few exceptions, all commands must be preceded by a backslash \.

For noninteractive use you must specify a source file, `article.tex`, for example. At the operating system prompt enter

```
C> tex article
```

Note that the `.tex` extension need not be entered. If the extension is not `.tex`, then you must type in the full name of the file. TeX will process the file, give the appropriate messages, and write the output and log files, just as in interactive mode. You will not get the `*` or `**`-prompts. (You will get the `*`-prompt after your source file is processed if it does not contain the `\bye` command. In this case you are back in interactive mode and you may continue to enter additional text and commands. To finish, enter `\bye`.)

If you enter an invalid filename or if the file does not exist, you will get the prompt

```
I cannot find file 'chatper1.tex'
```

```
Please type another input file name:
```

where '`chatper1.tex`' is the name of the file you specified but TeX did not find, because of the misspelling. You now have three options. You can enter the correct name, in which case everything will proceed as described above. You can enter `nul`, which cancels the request to input a file and leaves you at the *-prompt (see also §1.7 on input files). Finally, under VTeX, you can enter **Ctrl-Z** (hold down the **Ctrl** key and press **Z**). The last option is an *emergency stop*. It immediately aborts all processing and returns you to the operating system. **Ctrl-Z** can be typed whenever TeX expects terminal input. Use it only as a measure of last resort since some of your work may be lost.

If you type a command (anything that starts with the backslash '\') at either the **-prompt or *-prompt, the command is executed in the same way. However, if what you type is not a command, at the *-prompt it will be processed as input, while at the **-prompt it will be viewed as the name of an input file.

It is possible to typeset simple documents without knowing any commands (except `\bye`, which must be entered at the end of your file). TeX will automatically left- and right-justify your copy, do kerning and proportional spacing, paginate and typeset the output in the default font (roman).

The `\bye` command is the most important command because every source file or TeX session must be concluded with it. `\bye` tells TeX to ignore any subsequent text or commands and causes it to finish processing and exit to DOS. (If you omit `\bye`, TeX will think that there is more input on the way and will *-prompt you for it.)

The `\end` command may be used instead of `\bye`.

The file `sample1.tex`, is a simple document that contains exactly one command: `\bye`. By typesetting it you will see how text is formatted and learn a few basic rules.

Here are the contents of `sample1.tex`:

```
This  small  sample  file   shows  what  you can do with
TeX   without knowing any  commands. To be  precise, you
have  to  know exactly one  command—but it cannot  be
typed at  this          place in the file.

                                              If you type a
backslash, followed by b, y, and e in
order, to refer to the
command "by alias",
```

```
TeX will think it has reached the end of the
file and will not process any text or commands
that follow.
This may lead
you to believe that you can    never insert a
backslash in your document  because
TeX will expect a command
after     it. Later in this book   you   will find a way
to solve this problem.

Notice also that double quotes are entered as two single
opening and closing quotes and not by using
the double-quote key.
\bye
```

To typeset the file, prepare this file with your text editor. Then, at the operating system prompt, enter

```
C> tex sample1
```

First you will see the copyright notice and version information and then

```
**sample1
```

which is just a reminder that the job name (filename of the main source file) is "`sample1`". Next, you will see

```
(sample1.tex
```

The *opening* parenthesis '(' tells you that TeX opened `sample1.tex` and is working on it. Shortly afterwards the line becomes

```
(sample1.tex [1])
```

Here `[1]` stands for page 1 of the copy just processed. If the document contains more than one page, the page numbers will be displayed as the pages are processed. In this example the whole copy contains only one page and the *closing* parenthesis ')' signifies that TeX finished processing `sample1.tex`. TeX then displays the message

```
Output written on sample1.dvi (1 page, 1124 bytes).
Transcript written on sample1.log.
```

and returns you to the system prompt.

The resulting file can be printed or previewed using an appropriate printer or display driver. This book does not discuss the specifics of the printer or display driver operation. You should refer to your implementation documentation.

This is how the final output of `sample1.tex` should look:

> This small sample file shows what you can do with TeX without knowing any commands. To be precise, you have to know exactly one command—but it cannot be typed at this place in the file.
>
> If you type a backslash, followed by b, y, and e in order, to refer to the command "by alias", TeX will think it has reached the end of the file and will not process any text or commands that follow. This may lead you to believe that you can never insert a backslash in your document because TeX will expect a command after it. Later in this book you will find a way to solve this problem.
>
> Notice also that double quotes are entered as two single opening and closing quotes and not by using the double-quote key.

If you examine this output you will notice the following:

- Paragraph breaks occurred where there were blank lines in the source file.
- Line breaks (carriage returns) and consecutive spaces in the input file have no effect on the output.
- The copy is right-justified and the first lines of paragraphs are indented.
- TEX distinguishes between opening (") and closing (") quotation marks, as in "by alias."
- The space between words is not evenly distributed. More space is left after commas and still more between sentences.
- Some letters are kerned for better appearance: compare Way with Way (Kerned letters are letters which are typeset closer together than usual in order to decrease the amount of white space. In the first "Way" above, the "W" and the "y" overhang the middle "a"; in the second they do not.). Kerning is more noticeable when large print is used. Example below is set in the `mvtmsr` and the `mvfutr` fonts.

  WayWay — WayWay

- TEX knows about ligatures: compare file with file. (A ligature is a group of letters connected in a special way for better visual effect. In the first file above, the crossbar of the "f" and the top of the "i" are connected and the top curl of the "f" serves as the dot of the "i". In the second file these letters are not connected.). Like kerning, ligatures are more noticeable in a large print:

  file *file* — file *file*

1.5 More Basics

TOPICS ▷ \TeX, \centerline, special characters

Here are a few more things you should know before you try TeX on your own:

- You can use any text editor to create a TeX source file. Do not use word processors that save files in other than ASCII format because TeX will not process such files correctly.
- TeX uses the backslash '\' character to distinguish between text and commands. (TeX users call this character *the escape*; do not confuse it with the ASCII escape character chr(27).) Characters that immediately follow the backslash form a command. Built-in command names (as opposed to user-defined ones) must be blocks of letters without embedded punctuation marks or spaces (\centerline, not \center line, for example), or they must consist solely of non-letter symbols (\! or \$, for example).
- Certain other characters are also used to give special instructions. They are: percent sign '%', left brace '{', right brace '}', dollar sign '$', number sign '#', less than '<', greater than '>', ampersand '&', underscore '_', caret '^' and tilde '~'. If one of them is used in a document, TeX will treat it as a command, rather than typeset it. If these characters need to appear in the output, then the following control sequences must be used to enter them:

Reserved Characters		
Character	Meaning in TeX, see	Control Sequence
%	comment char, §1.5	\%
$	math switch, §4.1	\$
{	group opening, §2.4	\{
}	group closing, §2.4	\}
#	macro argument, §6.2	\#
&	tab position,§4.12,§9.1	\&
_	subscript, §4.2	_
^	superscript, §4.2	\^
~	tie-in, §3.9	\~
<	*none*	<
>	*none*	>
\	escape char, §1.4	\backslash

The \{, \}, and \backslash commands may be used only in math mode. While the < and > characters are not commands, many traditional TeX fonts do not contain these symbols. For example, in Computer Modern fonts, Spanish question and exclamation marks occupy the positions normally reserved for < and >. Thus, < should be typeset in math mode: $<$. With VTEX, these characters work as should.

EXAMPLE 1.1 ▷ In order to typeset

$5 worth of #4 nails at 30% off

you must type

```
\$5 worth of \#4 nails at 30\%\ off
```

- Spaces after a command are ignored. If you type '`\TeX  is easy`', it will be typeset as TEXis easy, with no space after "TEX". To provide a space, type '`\TeX\␣is easy`' instead. Similarly, '`\$␣5`' is equivalent to '`\$5`'. You should type '`\$\␣5`' to obtain $ 5. '`\␣`' (backslash and space) is called the *visible space command*; it inserts a space in the text.
- The example in the previous section had a small flaw in it: instead of TEX, it said TeX. The '`\TeX`' command will typeset the logo as it should be – with "E" lowered. Note that all commands are *case-sensitive*, that is, '`\TeX`' is not equivalent to '`\Tex`' or '`\TEX`'. The logo is corrected in the file `sample2.tex`, which is otherwise identical to `sample1.tex`. To typeset it, proceed as above, but use the file `sample2.tex`.
- Multiple spaces are also ignored. Typing

```
                    My first \TeX\ file
```

  will not create a centered title; rather, the line will be indented and justified:

My first TEX file

In order to center a line, you must use the `\centerline` command:

```
\centerline{My first \TeX\ file}
```

which produces

My first TEX file

Notice that the left '`{`' and right '`}`' braces are used to mark the beginning and the end of the text to be centered. If the braces are omitted, the output will be

M
y first TEX file

Notice that only the first character is centered and that the line was split after it.

- The % character is used to start a comment. Anything between a % and the end of the line will not be processed. For example:

```
% Term paper, last revision 04/01
Columbus discovered America in 148?. % Check the date!
```

Only the words "Columbus discovered America in 148?." will appear in the output. Comments can be used to keep track of revisions of the document, to mark intended changes, or to temporarily exclude parts of the text.

The use of the other special characters will be discussed later.

TEX has hundreds of commands designed to handle almost any application. However, it is not necessary to remember all or even most of them. The commands described so far are sufficient for typesetting many simple documents.

1.6 Introduction to Error Handling

Error handling in TEX is best illustrated with an example. The file sample3.tex, shown below, contains some errors. Here are its contents:

```
This small file demonstrates
  error handling in \TEX\
We just misspelled a command and now we will
forget an end.
```

In this file, \TEX is an invalid command (recall that commands are case-sensitive) and \bye is missing at the end of the file. To process this file type

```
C> tex sample3
```

After the usual displays you will see the following:

```
! Undefined control sequence.
l.2    error handling in \TEX
                             \

?
```

First comes the error message, which is usually self-explanatory. Next, TEX shows the line that contains the error (Line 2 of the input file) and splits it at the word that caused the problem. Finally, the question mark (?) prompts you for instructions.

Type a question mark to obtain the list of options:

```
? ?
Type <return> to proceed, S to scroll future error messages,
```

```
R to run without stopping, Q to run quietly,
I to insert something, E to edit your file,
1 or ... or 9 to ignore the next 1 to 9 tokens of input,
H for help, X to quit.
?
```

The more important options are:

- <return> (i.e. pressing the return or enter key) lets TeX handle the error itself. In this example, it would ignore the undefined control sequence \TEX and go ahead as if it were not there. The output would lack a word, but the error would be recorded in the log file sample3.log. By using this file you would be able to correct the error permanently.
- The 'Q' option (quiet run) tells TeX to correct this and any subsequent errors and record them in the log file for later reference, without prompting you for instructions.
- The 'X' option (quit) returns you to DOS – it is the last resort.
- Typing 'H' always produces an explanation of the nature of the error; sometimes you will even get a suggestion. In this example 'H' produces:

```
? H
The control sequence at the end of the top line
of your error message was never \def'ed. If you have
misspelled it (e.g., '\hobx'), type 'I' and the correct
spelling (e.g., 'I\hbox'). Otherwise just continue,
and I'll forget about whatever was undefined.
?
```

In the above message, "\def'ed" refers to the \def macro definition command that will be explained in §6.1. Typing I\TeX solves the problem (but only for now: the spelling in the sample3.tex sample file has not been changed).

Next you will see a closing parenthesis ')', indicating that the reading of sample3.tex is finished and the file is closed. However, instead of returning to the operating system you will get the *-prompt. The reason is that the '\bye' command was not encountered and TeX is waiting for more input. You can add a few more lines of text, enter a command or even input another file (see the \input command in the next section); for now just type \bye and return to the operating system.

1.7 Input Files

TOPICS ▷ `\input`, `\endinput`

As mentioned before, input can be read from several files. The `\input` command tells TeX to read another file. The files loaded by this command will sometimes be called *include files*. A file may actually contain only `\input` commands (and an `\end`, of course):

```
% This is the main file of the document.

\input book.sty
\input chapter1
\input chapter2
\input chapter3
\input appendix
\input index
\end
```

The files `chapter1.tex`, `chapter2.tex`, `chapter3.tex`, `appendix.tex`, and `index.tex` contain chapters 1-3 of the document, the appendix, and the index, respectively. The file `book.sty` is a stylesheet for book typesetting. It defines parameters to be used in typesetting a book and it may also define additional typesetting commands that are useful for typesetting whole books. `\input` can be used to keep a document from becoming uncomfortably large. Portions of a document may be typeset separately and brought together only when creating the final copy. Include files may themselves read in other files with the `\input` command. TeX allows up to seven levels of input files.

The `\input` command should always be followed by a valid filename; if the file extension is missing, `.tex` will be assumed. As before, if the file does not exist, you will get the prompt

```
Please type another input file name:
```

and the same three options are available: entering the correct file name, `nul`, or `Ctrl-Z`. If you enter '`nul`', processing of the file which contains the `\input` command resumes and you do not get the *-prompt.

The `\endinput` command is similar to `\end`, except that it is used only in include files. It terminates processing of an include file prior to its end. `\endinput` is not always required; if it is omitted, TeX will read to the physical end of the file and then return to the file that was previously being processed.

Chapter 2

Fonts & Characters

2.1 Typesetting: A Historical Overview

During the first few centuries after typesetting was invented, it remained a kind of semi-secret art. Rather than being the result of a conspiracy, this secretiveness was a natural consequence of highly specialized and difficult-to-learn techniques not available to the uninitiated. Typesetters formed independent guilds, with knowledge and tricks passed from masters to pupils, often from a father to his son. Centuries of isolation is a breeding ground of dialects, and typesetters, being no exception to the rule, naturally developed their own jargon. Since TeX, like any other modern typesetting system, borrows from these centuries-old traditions, it is helpful to review some of the history of typesetting and introduce some of the traditional terminology.

The earliest printed books date back to the 13th century. The very first books were set using wooden matrices for entire pages, but by the late 14th century Koreans and Chinese started using metal type. By the middle of the 15th century Europeans invented movable type. It is probable that the credit for this lies with the German printer Johann Gutenberg, who demonstrated his movable-type machine in 1448; it is certain that by the end of the 15th century typesetting had spread throughout Europe.

The technology of typesetting remained nearly unchanged for almost four centuries. And during this time, thousands of typesetters designed and experimented with different typefaces. The design of a typeface is more of an art than a science, and like any other type of art it went through many phases.

The earliest typefaces were mostly variations of Fraktur or Gothic, since these were the typefaces most similar to the contemporary written manuscripts. The Cloister typeface is a modern variation on such a design:

Cloister: A B C D E F a b c d e f

While a modern typesetter would use this design sparingly, and mostly for titles, at the dawn of typesetting, typefaces like Cloister were used to typeset entire books.

Roman typefaces were introduced toward the end of the 15th century. However, the name came to be used much later (Philippe Grandjean, *romain du roi Louis XIV*, 1693). The first italic design dates back to 1501 (Aldus Manutius). Original roman typefaces are now known as *Old Style Roman*:

Garamond: A B C D E F a b c d e f
Goudy: A B C D E F a b c d e f

The common feature of the Old Style designs is lack of emphasis; most strokes tend to be of approximately the same width, and the serifs are often of the same width as the stems. It took a refinement of technology (better paper, better ink, and better printing techniques) before the appearance of first *Transitional* and, later, *Modern* typefaces:

Baskerville: A B C D E F a b c d e f
Bodoni: A B C D E F a b c d e f

Baskerville (John Baskerville, circa 1750) is perhaps the most successful Transitional typeface; it displays a combination of thin and thick strokes. Modern typefaces, like Bodoni (Giambattista Bodoni, late 1700's), take this approach to the extreme.

Many of the contemporary typefaces were designed at the end of 1800's or early 1900's, when the advent of new mass-printing equipment (Linotype, Otto Mergenthaler, 1884) and the emergence of mass, and quality, printed magazines created markets for new typefaces. Two common examples are Times and Century:

Times: A B C D E F a b c d e f
Century: A B C D E F a b c d e f

Both Times and Century were commisioned for specific publications: Century for the Century Magazine (L. B. Benton and T. L. DeVinne, 1894), Times for the London Times (S. Morison, 1931). These are two immensely popular designs and as such exist in many variations.

Sans-serif typefaces are relatively new and did not become widely used until the 20th century. Helvetica, the most popular sans-serif design, was introduced only in 1957. Franklyn Gothic (M. Benton, late 1900's) is one of the oldest sans-serif designs:

Helvetica: A B C D E F a b c d e f
Franklin Gothic: A B C D E F a b c d e f

TeX was designed in the late 1970's when quality typefaces were not readily available in computer-readable form. It was therefore necessary to design an entirely new family of typefaces, suitable for printing on a variety of computer-driven equipment. The end result of a multi-year effort was the Computer Modern (CM) font family which to this day plays a central role in TeX. The Computer Modern family contains three primary text fonts: Roman, Sans, and TeleType. Of the three, Roman is a fairly weak clone of the Century SchoolBook:

CM Roman: A B C D E F a b c d e f
Century: A B C D E F a b c d e f

Sans and TeleType are, however, independent designs; TeleType also happens to be one of the best designs of a non-proportional typeface:

CM Sans Serif: A B C D E F a b c d e f

`CM TeleType: A B C D E F a b c d e f`

From the beginning, Computer Modern fonts provided both help and damnation to TeX. While its very presence was instrumental in porting TeX to all kinds of computers, it turned many a potential user away from TeX.

Today, the strength of the family lies in its completeness: Computer Modern fonts include a matched set of mathematical fonts, something that most non-TeX fonts lack. This effectively assures the survival of CM text fonts, since (better) non-CM fonts may not have their own mathematical symbols and often would not blend well with CM math fonts.

This book has been typeset in a combination of CM and non-CM fonts. The text and titles are set in `mvmelr` and `mvfutb`; these are VTeX fonts that are somewhat related to Melior and Futura. However, we used CM TeleType for setting pieces of the `entered text`; the typeset examples are mostly set in CM Roman, since this is the usual default font in TeX.

2.2 Font Representation

While traditional typefaces were stored as a *metal type*, computerized typesetting relies on fonts stored as computer files. The most common form is the *bitmap representation*: the characters are stored as binary matrices of pixels; a bit of 1 represents a pixel that is turned on, while 0 represent a pixel that is turned off. Below is a sample bitmap representing the letter 'A':

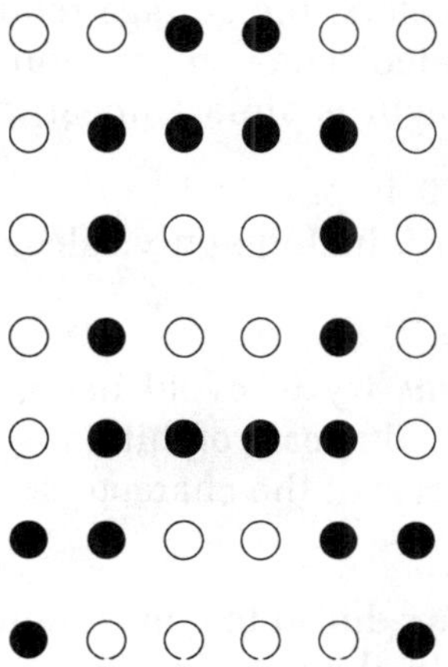

There are many of different file formats based on bitmap representation, both in TEX and outside it. Some common formats are

- LaserJet softfonts
- MicroSoft Windows Screen Fonts
- TEX `.PXL` (pixel) fonts
- TEX `.PK` (packed) fonts.

Despite differences in formats, all of these formats (and many others) are very similar and convertible one to the other. They also share the same advantages and disadvantages. On the positive side, bitmap fonts are very fast to use. Also, since bitmap fonts describe characters explicitly, up to the last dot, they can be hand-crafted for the best results. The main disadvantage of bitmaps is their storage requirements. Below is a table that displays the size of the bitmap font files:

Font (points)	PXL (bytes)	PK (bytes)
10	14916	5228
20	49592	10068
30	79336	15768
40	145768	23304
50	211412	30688

Observe that the `.PXL` files exhibit approximately a quadratic order of growth: when you double the size of the characters, the storage requirements quadruple. The `.PK` format utilizes a very sophisticated data-compression scheme to reduce the size of the files so that the order of growth is almost linear.

The use of bitmap fonts makes it impractical to maintain too many font files. Most of TEX implementations, therefore, include only selected font sizes: usually 5, 7, 10, 12, and 14 points.

Contemporary typesetting systems try to avoid using bitmap fonts because of their excessive storage requirements. Instead of bitmaps the fonts are represented by formulas, describing the boundaries of the characters. Immediate advantages of these *vector fonts* are:

- Storage requirements are reduced to almost nothing. For instance, the `MVTT10` typeface used in the table above takes about 23kb as a vector font.
- All sizes of the fonts become available, including very large ones.

- Many font effects become feasible. These include effects obtained by a coordinate transformation of the font-describing equation, such as slant, compression, or rotation.

As an example, traditional TEX systems include both Roman and *Slanted Roman* typefaces. A vector-based system, such as VTEX, does not need the *Slanted* typeface, since it can easily build it from its upright cousin.

Vector font formats include

- PostScript fonts
- True-Type fonts
- HP LaserJet III (PCL5) fonts
- VTEX vector fonts

Vector font representations are often based on different ideas. Therefore, conversion between different vector formats is often difficult or impossible. However, vector fonts share the same features in all formats:

- Low storage requirements
- Much greater flexibility, including support for font effects
- Greater processing time
- Somewhat worse quality, especially at low resolutions

2.3 Standard Fonts

TOPICS ▷ `\bf`, `\it`, `\rm`, `\sl`, `\tt`

Most typesetting jobs require only a few standard fonts. These fonts can be selected with short mnemonic commands.

The **boldface font** can be accessed by using the command `\bf`.

EXAMPLE 2.1 ▷ Typing

```
I'd like to switch to bold... \bf and now I did!
```

produces

I'd like to switch to bold... **and now I did!**

The '`\bf`' command works as a switch, telling TEX to start typesetting in **boldface**.

Similarly, you can use `\it` to invoke the *text italic* font (italic for short), `\sl` for the *slanted* font and `\tt` for the `typewriter font`. The default font (the font used if no font commands are specified) is the roman font. The command `\rm` can be used to select the roman font. The following example illustrates these fonts.

EXAMPLE 2.2 ▷ Typing

```
\rm A Quick Brown Fox Jumps Efficiently over the Lazy Dog.

\bf A Quick Brown Fox Jumps Efficiently over the Lazy Dog.

\it A Quick Brown Fox Jumps Efficiently over the Lazy Dog.

\sl A Quick Brown Fox Jumps Efficiently over the Lazy Dog.

\tt A Quick Brown Fox Jumps Efficiently over the Lazy Dog.
```

results in

> The Quick Brown Fox Jumps Efficiently over the Lazy Dog.
> **The Quick Brown Fox Jumps Efficiently over the Lazy Dog.**
> *The Quick Brown Fox Jumps Efficiently over the Lazy Dog.*
> *The Quick Brown Fox Jumps Efficiently over the Lazy Dog.*
> `The Quick Brown Fox Jumps Efficiently over the Lazy Dog.`

Make sure you enter the blank lines in the input so that each sentence will appear on a separate line.

Notice that all these fonts, with the exception of the `typewriter` font, are proportional. The latter was designed to emulate teletypes, typewriters and computer screens, which lack proportional spacing. In non-proportional (or *monospace*) fonts all letters have the same width.

A reader familiar with "quick brown foxes" will recall that this sentence (less the word "Efficiently") is frequently used to test typewriters, since it contains all 26 letters of the alphabet. "Efficiently" was added to demonstrate another feature of TEX fonts, namely ligatures. For example, in italics '`ffi`' is typeset as [*ffi*] and not as [*ffi*]. The other standard ligatures are [*ff*], [*fi*], [*fl*] and [*ffl*]. Ligatures are not formed in monospaced fonts, like the `typewriter` font; they are also omitted in the SMALL CAPS fonts. In most sans-serif fonts the [ffi] ligature is formed by reshaping the second [f] instead of combining all three letters.

The most commonly used CM fonts are roman, **bold** and *italics*. This book uses the MicroPress Meller family of fonts to avoid confusion with the examples. The bulk of the book is typeset in Meller Roman, with *Meller Italics* and **Meller Bold** used mainly for emphasis.

Many people (and many computer manuals) view *italics* as "opposite" to upright. In fact, the *italic* font is not only slanted, but it also uses character shapes that are more script-like than roman. Moreover, italics were designed originally as independent fonts, not as counterparts to roman typefaces. Since there also exist unslanted italic fonts, it is the *slanted* font that may be viewed as being "opposite" to roman, since its characters have the same basic shape as in roman, the only difference being the *slant*.

There are many fonts in TEX, but these five are sufficient for most ordinary uses. The following example shows how fonts can be mixed. It also shows how font mixing can be overdone.

EXAMPLE 2.3 ▷ Typing

```
One of the first things that \sf new users \rm do
is to try to use as many fonts as possible.
    They mix \bf boldface
\rm and  \sl slant  \rm with  \it italics  \rm and
 add \tt teletype  for more spice.
If they forget to change the font back to roman, the rest
of the document will be typeset in the last selected font.
```

produces

> One of the first things that new users do is to try to use as many fonts as possible. They mix **boldface** and *slant* with *italics* and add `teletype for more spice. If they forget to change the font back to roman, the rest of the document will be typeset in the last selected font.`

All font commands work as switches. Combining font commands will not produce a hybrid font. Thus, typing '`\bf\sl`' within the text will not produce a bold slanted font. It will simply select the *slanted* font, since '`\sl`' was specified last.

Most documents are printed using only a few carefully chosen fonts. Using every available font on a page is fine when typesetting a catalog of fonts or a book such as this one but a business letter printed this way is more likely to annoy than to please. The choice of fonts is usually a matter of personal taste, but the guidelines below are generally followed.

- Be consistent. Use one font for the body of the text, one font for section headers, one font for running heads, and one font for quotations.
- Try not to use similar fonts in the same document. For example, *mixing italic letters with slanted letters* produces unattractive results.

- For optimum readability, use the roman font or the sans-serif font to typeset the document. Reading long documents typeset in boldface or (unslanted) italics can tire the eyes. Italics should be used to typeset at most a few paragraphs.
- Use **boldface** for emphasis and in headers.
- Use CAPS AND SMALL CAPS for initial lines and for personal names (some books do this).
- Use *italics* or *slant* for emphasis or to typeset quotations and references.
- Use the `typewriter` font for examples in computer manuals or program listings.

➡ When TEX determines the spacing between characters, it does all the computations with respect to the spacing on the *baseline*. The baseline is the line on which characters without descenders (e.g. 'a','b' or 'c', but not 'g' or 'j') are sitting. When switching from a font slanted right to an unslanted font or to a font slanted left, the resulting spacing might be too small. The *italic correction* command '\/' may be used to adjust the spacing.

EXAMPLE 2.4 ▷ Typing

```
This is how text looks
\it without \rm the correction and \it
 with\rm\/ the correction.
```

results in

This is how text looks *without* the correction and *with* the correction.

➡ The italic correction is used *after* the italic and the slanted fonts, as well as *before* certain fonts slanted left, such as the funny font.

2.4 Groups

TOPICS ▷ `{ , }`

Usually, a document is typeset in a single font, with others being used only for small portions of the text. The main font (if different from the default roman) is specified by a font command at the beginning of the document. To return to the main font after a piece of text typeset in a different font, an explicit command is required.

EXAMPLE 2.5 ▷ Typing

```
This sentence has a \bf few bold words \rm in the middle.
```

results in

This sentence has a **few bold words** in the middle.

The command `\rm` is used here to return to the default roman font. The disadvantage of this method is that the user has to remember what the main font is. If the main font had been italics, for example, then the command `\it` would have been used to return to it. Moreover, if the main font of a document is roman and this font needs to be changed to italics, then every occurrence of '`\rm`' which restores the main font must be replaced by '`\it`'. This can be a tedious process.

To simplify font changes TeX uses *groups*. The left '`{`' and right '`}`' braces mark the beginning and the end of a group.

EXAMPLE 2.6 ▷ In

```
Jill felt {wonderful} about Jack.
```

the group contains the word "wonderful". In this case the braces have no effect on the output, which is

Jill felt wonderful about Jack.

Font commands inside a group affect only that group.

EXAMPLE 2.7 ▷ Typing

```
Jill felt {\it wonderful}\/ about Jack.
```

produces

Jill felt *wonderful* about Jack.

Whenever TeX sees the beginning of a group, it notes the current font and restores it after the end of the group. In the example above, the word "about" is typeset in roman because that font was in effect before the group was opened. The same result would have been obtained by typing

```
Jill felt \it wonderful \rm about Jack.
```

The font change command does not have to appear at the beginning of the group.

EXAMPLE 2.8 ▷ Typing

```
Jill {felt \it wonderful} about Jack.
```

produces

Jill felt *wonderful* about Jack.

The '\it' command affected only the rest of the group.

Groups can be nested and a group may span several paragraphs.

EXAMPLE 2.9 ▷ The following text

```
Starting  with  the  default  roman  font,
{open a group, \it switch to italics,
{\bf open another group and switch immediately to

boldface},
close the inner group and revert to italics}
and finally return to the roman we started with.
```

will be typeset as

Starting with the default roman font, open a group, *switch to italics,* **open another group and switch immediately to** **boldface**, *close the inner group and revert to italics* and finally return to the roman we started with.

2.5 Underlining

TOPICS ▷ \underbar, \hrule

Underlining is used on typewriters, where the lack of fonts makes it the only way to emphasize text. However, underlining only rarely appears in professionally typeset documents. Underlining in TEX is done with the \underbar command.

EXAMPLE 2.10 ▷ Typing

```
The \underbar{text to be underlined}
 must be enclosed in braces.
```

results in

The text to be underlined must be enclosed in braces.

Notice that the `\underbar` command must come before the text to be underlined. If the braces are omitted, then only the character following the command will be underlined.

EXAMPLE 2.11 ▷ Typing

```
Proper nouns should be \underbar capitalized.
```

produces

Proper nouns should be capitalized.

The difference between `\underbar` and font commands is the following: font commands are switches which affect the rest of the text, while `\underbar` affects only the next character. If this character is a left brace, `\underbar` will find the matching right brace and it will underline the group delimited by the two braces. If braces are omitted, font commands will do too much, while `\underbar` will do too little.

The `\centerline` command introduced in §1.4 works in a similar fashion: it affects either the next character or a group within braces. Many other commands have the same syntax.

The `\underbar` command can be symbolically described as

SYNTAX ▷ `\underbar` ⟨argument⟩,

where ⟨argument⟩ stands for a single character or a group.

Underlining can be combined with font changes.

EXAMPLE 2.12 ▷ Typing

```
Why \bf would \underbar{anyone \it
in the world \rm do} this?
```

produces

Why **would anyone *in the world*** do **this?**

The first word was typeset in the default roman, then the current font became boldface. TEX remembered this font when it opened the group after '`\underbar`'. The word "anyone" is still in boldface because no other font change occurred before it. The font changes inside the group to be underlined extend only to the right brace, where the boldface font again becomes current.

➡ Underlining in TeX is not intelligent; it ignores (and crosses) the descenders. This is partially because TeX does not really see the characters it typesets; TeX only knows the dimensions of the letters.

➡ `\underbar` is a build-in or "primitive" command in TeX. However, a knowledgeable user can design versions of `\underbar` that use broken or shaded pattern, see §8.10.

The `\underbar` command supplies a horizontal line or *rule* under the affected text. The `\hrule` command can be used to draw a horizontal rule across the page to separate segments of text.

SYNTAX ▷ `\hrule`

EXAMPLE 2.13 ▷ Typing

```
It is a common practice to separate sections of text
with horizontal rules. The rule is usually surrounded by
vertical skips, covered later in the book.
\hrule
You can also draw vertical rules
and solid black rectangles.
These are described later in the book.
```

results in

> It is a common practice to separate sections of text with horizontal rules. The rule is usually surrounded by vertical skips, covered later in the book.
>
> ---
>
> You can also draw vertical rules and solid black rectangles. These are described later in the book.

You can also specify the dimensions of a rule explicitly (see §7.2).

2.6 Other Fonts

TOPICS ▷ `\font`

TeX provides only five predefined font commands. If fonts other than the ones previously discussed are required, then new font commands may be defined or the meaning of the five pre-defined font commands may be changed. Defining or redefining font commands is done by using the `\font` command.

SYNTAX ▷ `\font` ⟨font command⟩ `[=]` ⟨font name⟩ ⟨options⟩

The `\font` command accepts two main arguments. ⟨font command⟩ is the user-given name of the font being defined and ⟨font name⟩ is the "official" name of the font that is assigned to ⟨font command⟩. ⟨font command⟩ may only contain characters and not digits or other non-characters.

EXAMPLE 2.14 ▷

```
\font\sc       =  cmcsc10
\font\sf       =   cmss10
\font\times =   mvtmsr
\font\helv        mvhlvr
\font\cloister  mvcloi
```

The equal sign (=) is optional as with most TEX commands and may be replaced with a space.

⟨font command⟩ usually is just the name of an appropriate *font metric file*. In most versions of TEX, the font metric file is an individual file with extension `.tfm`. VTEX allows metric files to be combined into a global font library. The example above will work if your implementation of TEX includes the needed `.tfm` files.

Naturally, different installations of TEX may have widely different selections of fonts. Some common standards are the Computer Modern fonts and Almost Computer Modern (CM and AM families), originally designed for use with TEX, PostScript fonts used for printing on PostScript devices, and VTEX vector fonts. The names of Computer Modern Fonts usually start with letters `cm`, Almost Computer Modern fonts start with `am`, and VTEX fonts with `mv` (MicroPress Vector). (See the appendix of this book for a list of some additional fonts.)

`\font` only needs to be issued once, and a font command must be declared prior to using the command. After declaring the font command it can be used in the same way that the five pre-defined TEX font commands are used. For instance:

EXAMPLE 2.15 ▷ Typing

```
\font\sans     = cmss10
\font\brush    = mvbrsh
\sans  We are using the new font we just declared.

\brush And now we're using the second font.
```

produces

We are using the new font we just declared. *And now we're using the second font.*

While writing a document it may not be clear what font will look best. Defining font commands makes it easy to switch fonts. For example, to change all the occurrences of the `typewriter` font in this book to Courier, it would be sufficient to place

```
\font\tt=mvcour
```

at the beginning of the source file. Note that this would make the typewriter font inaccessible. To use it, we will now have to define a new font command.

A new TeX user should try to print the fonts in his installation. It will allow him or her to ascertain

- what fonts are available
- which fonts look right on the available output devices (some may not)
- which fonts look right at which size
- what characters are present in what fonts

One of the ways to experiment is to print the same sentence (or a paragraph) in many fonts. The `showfont.tex` program (see §11.10) gives a more efficient solution to this undertaking, but for now we will simply enter the text by hand.

EXAMPLE 2.16 ▷ Typing

```
\font\cmu=cmu10      \font\cmsy=cmsy10  \font\cmmi=cmmi10
\font\helv=mvhlvr    \font\ding=mvding2 \font\cloi=mvcloi
    % Multiple commands may be specified on the same line

\cmu  A Quick Brown Fox Jumps over the Lazy Dog for \$10.

\cmmi A Quick Brown Fox Jumps over the Lazy Dog for \$10.

\cmsy A Quick Brown Fox Jumps over the Lazy Dog for \$10.

\helv A Quick Brown Fox Jumps over the Lazy Dog for \$10.

\ding A Quick Brown Fox Jumps over the Lazy Dog for \$10.

\cloi A Quick Brown Fox Jumps over the Lazy Dog for \$10.
```

results in

This example includes three Computer Modern fonts (cmu10, cmmi10 and cmsy10) and three MicroPress Vector fonts (mvhlvr, mvding2 and mvcloi). The output demonstrates that:

- of the six fonts, only `cmu10` and `mvhlvr` can be used for writing text
- `mvcloi` is a display font; it can be used for headlines, but not for ordinary text
- `cmu10` lacks a dollar sign character—a layout bug in some of the Computer Modern fonts
- `cmmi10` is the *Math Italics* font. The lack of spacing is due to the fact that this font is used for typesetting mathematics, where the spacing rules are different from those for ordinary text. Furthermore, math fonts have no ligatures – typesetting '$x = fi + j$' as '$x =$ﬁ$+j$' would be inappropriate
- `cmsy10` is also a math font, but it contains only symbols and calligraphic capital letters. (`cmmi10` also contains some symbols – for example, it has a triangle instead of the period)
- `mvding2` is a "pi" or "dingbat" font—it has no letters, just typesetting symbols. It is also a larger font than the rest

The following table lists some of the common Computer Modern fonts, as well as the names of disk files which contain them. (See §11.5 for a similar list of MicroPress vectors.)

Font	Command	Disk file
Boldface	`\bf`	`cmbx10`
CAPS AND SMALL CAPS	†	`cmcsc10`
Math Italics	†	`cmmi10`
MATH SYMBOLS	†	`cmsy10`
Roman	`\rm`	`cmr10`
Sans-Serif	†	`cmss10`
Slanted	`\sl`	`cmsl10`
Slanted Typewriter	†	`cmsltt10`
Text Italics	`\it`	`cmti10`
Typewriter	`\tt`	`cmtt10`
Unslanted Italics	†	`cmu10`

In Plain TEX, there are no predefined commands for fonts marked with †. LATEX and VTEX also define `\sf` and `\sc` for the Sans-Serif font and the Caps and Small Caps font.

2.7 Font Sizes and Magnification

TOPICS ▷ `at`, `scaled`, `\magstep`, `\magstephalf`, `\magnification`

All the font files listed so far have a '10' in their names. It indicates that these fonts have a *design size* of 10 points. The point is a unit of length used by typographers. An inch has about 72 points (72.27, to be exact).

While most books and newspapers are typeset at 10 points, you are not limited to this size. Fonts of different sizes are used in the following cases, for instance:

- Larger fonts for headers, titles, or running heads
- Smaller fonts for text in footnotes
- Smaller fonts for subscripts and superscripts

In some cases the entire document may be typeset in a size different from 10 point (see the next section).

The `\font` command discussed in the previous section can be used to load fonts designed in different sizes. CM typefaces include variant designs for selected point sizes. For instance, CM family has not just `cmr10` but also `cmr5`, `cmr6`, `cmr7`, `cmr8`, `cmr9`, `cmr12`, and `cmr17`.

EXAMPLE 2.17 ▷ Typing

```
\font\twelverm=cmr12
\font\tenrm   =cmr10
\font\ninerm  =cmr9
\font\eightrm =cmr8
\font\sevenrm =cmr7
\font\sixrm   =cmr6
{\twelverm twelverm} {\tenrm tenrm} {\ninerm ninerm}
{\eightrm eightrm} {\sevenrm sevenrm} {\sixrm sixrm}
```

produces

twelverm tenrm ninerm eightrm sevenrm sixrm

➡ As mentioned above, '`\font\rm6=cmr6`' would be wrong since names of font commands may not contain digits.

A careful examination of the output shows that the fiverm font is not the same as the tenrm font reduced by a factor of two – the shapes of the characters are somewhat different. To see the difference, you use the `\font` command to magnify the `cmr5` font to 10 points.

EXAMPLE 2.18 ▷ Typing

```
\font\twicefiverm=cmr5 at 10pt  \twicefiverm Font Sample

\rm Font Sample
```

produces

Font Sample. Font Sample.

Clearly, these two fonts are different.

➡ Computer Modern is the only font family that has different font designs for different point sizes. While there are certain advantages to have them (namely, better subscript fonts), there is also a price for carrying along a very large number of font variants. The multiple matrix fonts from Adobe Systems, or use of VTEX's `aspect` and `bold` options offer much more economical solutions.

Scaling may be used to produce fonts of different sizes. There are two ways to do scaling with the `\font` command.

SYNTAX ▷ `\font` ⟨font command⟩ = ⟨font file name⟩ `at` ⟨desired point size⟩

In the example above, the `\twicefiverm` command was defined in order to access the `cmr5` font scaled at 10 points.

SYNTAX ▷ `\font` ⟨font command⟩ = ⟨font file name⟩ `scaled` ⟨1000×scale-factor⟩

Only the result of the multiplication should be entered, not the multiplication itself.

EXAMPLE 2.19 ▷ An alternative way to define `\twicefiverm` is to specify

```
\font\twicefiverm=cmr5 scaled 2000
```

EXAMPLE 2.20 ▷ To define a half-size `cmr10`, use either

```
\font\halfcmr10=cmr10 scaled 500
```

or

```
\font\halfcmr10=cmr10 at 5pt
```

Remember, this font is different from `cmr5`.

TEX pays special attention to scaling fonts by a factor of 1.2 and uses the `\magstep 1` command to specify it.

EXAMPLE 2.21 ▷ Specifying

```
\font\bigrm=cmr10 scaled 1200
```

is equivalent to

```
\font\bigrm=cmr10 scaled \magstep 1
```

Similarly, `\magstep 2` denotes scaling by a factor of $1.44 = 1.2 \times 1.2$, `magstep 3` scales by a factor of $1.728 = 1.2 \times 1.2 \times 1.2$, etc. You may also use the `\magstephalf` command to scale by a factor of $\sqrt{1.2} \approx 1.095445$.

CAUTION ▷ TEX can do any kind of scaling, but in order to print the output you must either

- have the font in the correct size, or
- have a driver capable of smooth font scaling.

Check your driver documentation for more details on this point.

The `\magstep` command is primarily designed to support drivers that cannot carry out smooth scaling. These drivers are generally shipped with a collection of fonts for the default size and fonts for the sizes specified by `\magstep` 1, 2, and 3. For example, a driver for an Epson MX-series printer may be shipped with fonts designed for resolutions of 120, 144 ($= 120 \times 1.2$), 173 ($\approx 144 \times 1.2$) and 208 ($\approx 173 \times 1.2$) dots per inch.

When the entire document needs to be typeset in a different point size, the `\magnification` command should be used. This command scales not only all the fonts but all the dimensions as well.

SYNTAX ▷ `\magnification` [=] ⟨1000 × desired scaling factor⟩

where the equal sign '=' is optional. Only the result of the multiplication should be entered, not the multiplication itself.

The `\magnification` command may only be used once, at the beginning of the document.

EXAMPLE 2.22 ▷ Specifying

```
\magnification=2000
```

will scale the entire document by a factor of two ($2000 = 2 \times 1000$). All 10-point fonts will appear at 20 points. If the spacing between lines (see §3.12) was 15 points, this command will change it to 30 points. However, since the physical dimensions of the page remain the same, all lines will contain fewer characters and all pages will contain fewer lines and more sheets will be needed to print the document. To scale the whole document by a factor of 1/2, specify

```
\magnification=500
```

because $500 = \frac{1}{2} \times 1000$. The command

```
\magnification=1000
```

will leave the document unchanged ($1000 = 1 \times 1000$).

➡ If the `\font` command is used to scale all the fonts in a document individually by a factor of 1.2, for example, the result will not be identical to the global `\magnification=1200`, because in the former case spacing is not rescaled.

Sometimes it is desirable to magnify the document without magnifying a particular font. TeX allows users to specify *true* dimensions. For instance, specifying

```
\font\alwaysten=cmr10 at 12truept
```

makes `\alwaysten` to appear always at ten points, regardless of the settings of the `\magnification`.

2.8 Font Attributes

TOPICS ▷ `aspect`, `slant`, `outline`, `shadow`, `fillpattern`, `smallcaps`
▷ `\Aspect`, `\Slant`, `\Outline`, `\Shadow`, `\Fillpattern`
▷ `\Smallcaps`, `\fontscale`.

VTeX allows a number of additional parameters to be specified with the `\font` command. These additional attributes allow the user to control font proportions and slant, as well as implement visual font effects such as outline and shadow.

- **Aspect Ratio.** `aspect`⟨n⟩ specifies the aspect ratio, or the amount of vertical stretching applied to the font. (1000 is the default, 500 gives a half-height font, 2000 gives a double-height font).

EXAMPLE 2.23 ▷ Typing

```
\font\t=mvpalr aspect 1200 \t This is a tall font. \par
\font\t=mvpalr aspect 1000 \t This is normal font. \par
\font\t=mvpalr aspect  800 \t This is a short font.
```

results in

> This is a tall font.
> This is normal font.
> This is a short font.

- **Slant.** `slant`⟨n⟩ specifies the amount of slant of the font (0 = upright font, 1000 = font slanted 45° to the right, -1000 = font slanted 45° to the left).

EXAMPLE 2.24 ▷ Typing

```
\font\t=mvkorr slant 250 \t This is slanted right. \par
\font\t=mvkorr slant 000 \t This is an upright font. \par
```

```
\font\t=mvkorr slant -250 \t This is slanted left.
```

results in

This is slanted right.
This is an upright font.
This is slanted left.

- **Small Caps.** `smallcaps` specifies a small caps font.

EXAMPLE 2.25 ▷ Typing

```
\font\t=mvr10  smallcaps
\t This is the automatically created small caps font.\par
\font\t=mvcsc10    \t This is the \TeX\ small caps font.\par
\font\t=mvr10   smallcaps slant   250
\t This is a slanted small caps font. \par
```

produces

THIS IS THE AUTOMATICALLY CREATED SMALL CAPS FONT.
THIS IS THE TEX SMALL CAPS FONT.
THIS IS A SLANTED SMALL CAPS FONT.

- **Outline.** `outline` specifies an outline font.

EXAMPLE 2.26 ▷ Typing

```
\font\t=mvtmsr scaled 3000 outline\t This is an outline.\par

\font\t=mvavnb scaled 3000 outline\t This is an outline.
```

produces

This is an outline.
This is an outline.

For best results, use `outline` only with font sizes of at least 20 points. For serif fonts, use outline only with font sizes of at least 30 points.

- **Shade or Fill Pattern.** `fillpattern`⟨n⟩ specifies the pattern used to fill a font. The value should be between 1 and 22 to select one of the predefined fill patterns.

EXAMPLE 2.27 ▷ Typing

```
\font\t=mvssbx10 scaled 3000 fillpattern 5
\t This is fillpattern \#5.   \par
\font\t=mvssbx10 scaled 2000 outline fillpattern 6
\t This is fillpattern \#6 with outline.
```

results in

This is fillpattern #5.
This is fillpattern #6 with outline.

- **Drop Shadow.** `shadow` specifies a font with a shadow.

EXAMPLE 2.28 ▷ Typing

```
\font\t=mvr10    scaled 3000 shadow \t This is a shadow. \par
\font\t=mvssbx10 scaled 3000 shadow \t This is a shadow.
```

results in

This is a shadow.
This is a shadow.

➡ For best results, use the `shadow` option in conjunction with either `outline` or `fillpattern`.

EXAMPLE 2.29 ▷ Typing

```
\font\t=mvr10    scaled 3000 shadow outline
   \t This is a shadow. \par
\font\t=mvssbx10 scaled 3000 shadow outline
   \t This is a shadow.
```

results in

This is a shadow.
This is a shadow.

- **Reverse Print & Backgrounds.** The extended form of the `fillpattern` option, `fillpattern "aabbcc` allows specification of the background pattern and reverse (white on black) print. The argument to the command is given as a 3-byte hexadecimal number (see §5.2.) The last byte (`cc`) determines the foreground pattern; `bb` determines the background pattern, while `aa` specifies the output operation:

  0 Normal Put

  1 Xor Put

  2 Or Put

  3 Nand Put

  4 Nxor Put

  5 Nor Put

  6 And Put

EXAMPLE 2.30 ▷ Typing

```
\font\t = mvhand fillpattern"10000 \t Reverse
\font\t = mvhlvx fillpattern"30300 \t Reverse
\font\t = mvhlvx fillpattern"10102 \t Reverse
\font\t = mvhlvx fillpattern"20201 \t Reverse
\font\t = mvhlvx fillpattern"10300 \t Reverse
```

produces

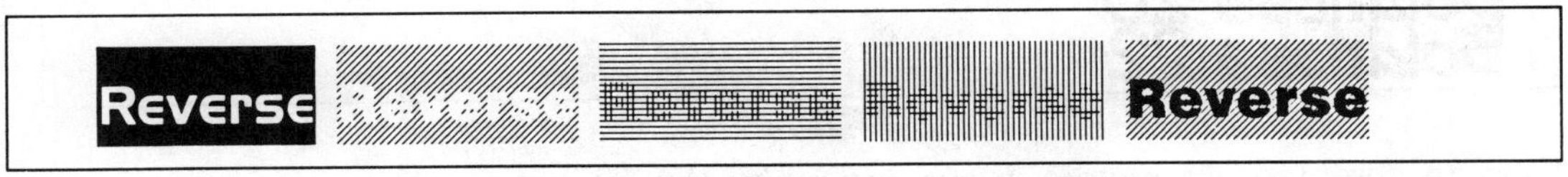

The height and depth of the background area are controlled with the `\bkgheight` and `\bkgdepth` commands. Specifying `\bkgheight 1000` and `\bkgdepth 1000` results in the areas shown above. Specifying `\bkgheight 500` and `\bkgdepth 2000` would double the background area below the baseline and reduce the area above the baseline by half.

- **Boldness.** `bold`⟨n⟩ varies the weight (or boldness) of the typeface. The `bold 0` selection corresponds to the normal weight; the positive values of ⟨n⟩ correspond to progressively heavier versions of the font, while the negative values will define progressively lighter versions.

EXAMPLE 2.31 ▷ Typing

```
\font\t = mvtekr scaled 2000 bold 00 \t Boldness 00\par
\font\t = mvtekr scaled 2000 bold 02 \t Boldness 02\par
\font\t = mvtekr scaled 2000 bold 04 \t Boldness 04\par
\font\t = mvtekr scaled 2000 bold 08 \t Boldness 08\par
\font\t = mvtekr scaled 2000 bold 10 \t Boldness 10\par
\font\t = mvtekr scaled 2000 bold 12 \t Boldness 12\par
\font\t = mvtekr scaled 2000 bold 14 \t Boldness 14\par
\font\t = mvtekr scaled 2000 bold 16 \t Boldness 16\par
\font\t = mvtekr scaled 2000 bold 18 \t Boldness 18\par
\font\t = mvtekr scaled 2000 bold 20 \t Boldness 20\par
\font\t = mvtekr scaled 2000 bold 22 \t Boldness 22\par
```

results in

Boldness 00
Boldness 02
Boldness 04
Boldness 08
Boldness 10
Boldness 12
Boldness 14
Boldness 16
Boldness 18
Boldness 20
Boldness 22

You can combine parameters in any way, with the following restrictions:

- The `scaled` and `at` parameters, if specified, should precede other parameters. In TEX-compatibility mode, VTEX ignores other parameters, treating them as text.
- If the same parameter is specified more than once (i.e. `slant 200 slant 230`), the last value is used.

EXAMPLE 2.32 ▷ Typing

```
\font\t=mvr10 scaled 1200 slant 250 \t Slanted and Scaled.
```

results in

Slanted and Scaled.

in normal mode, and produces

slant 250 Slanted and Scaled.

in compatibility mode. Notice that "slant 250" is printed in the font that was in effect before the new font declaration.

- **Alternative Font Effect Commands.** For each of the attributes described above, VTEX also has a corresponding command. The commands take exactly the same arguments as the attributes; they affect all the fonts in their range. These commands are:
  - `\Aspect{⟨n⟩}`
  - `\Slant{⟨n⟩}`
  - `\SmallCaps`
  - `\Shadow`
  - `\Outline`
  - `\Fillpattern`
  - `\Bold`
  - `\Fontscale{⟨n⟩}`

The last command corresponds to the `scaled` parameter of the font command. It allows local magnification of the text. (See §11.9 for more details.)

These commands are intended for local changes of font attributes that may involve several fonts. They are particularly helpful for customizing headers, titles, or footnotes.

➡ The `\Fillpattern` command affects not only fonts but also rules. This allows implementation of invisible fonts and color separation (see Dan Levin's article in the journal *TUGboat,* Vol. 13, #2).

2.9 Font Compatibility and Aliasing

✔ This section applies to VTEX only.

Different versions of TEX are not always completely compatible since they may use different fonts. For instance, some versions of TEX use the Almost Modern font family, while others use the Computer Modern family. If you have a document prepared with an AM-based system and you want to typeset it on a CM-based system, you generally should go through the source and replace all the references to the AM-family fonts with those in the CM-family. Similarly, since VTEX uses the MicroPress Vector font family, to process a document prepared for another system you should replace all the references to either CM- or AM- family fonts by those in the MV-family.

Font substitution can become a tedious process and the TEX standard should have provided means to automate it. VTEX enhances the TEX language with the `\aliasfont` command, which makes substitution transparent. The `\aliasfont` command is used to define entries in the internal font substitution table.

SYNTAX ▷ `\aliasfont` <old name> = <font spec>

where <old name> is the name of font being substituted and <font spec> is the name (with optional parameters) of the font that should be used instead.

EXAMPLE 2.33 ▷ Typing

```
\aliasfont cmr10 = mvr10
```

makes all references to the Computer Modern Roman font be interpreted as references to the MicroPress Vector Roman font. Without this `\aliasfont` command, typing

```
\font\rm=cmr10
```

would cause a 'font-not-found' error under VTEX, since VTEX does not use Computer Modern fonts. With `\aliasfont` in effect, VTEX will simply use `mvr10` instead of `cmr10`.

EXAMPLE 2.34 ▷ Typing

```
\aliasfont cmsl10 = mvr10 slant 333
```

makes all references to the Computer Modern Slanted font be interpreted as references to the MicroPress Vector Roman font, slanted by an appropriate amount.

The `\aliasfont` command is always global. A second occurrence of `\aliasfont` for the same font will override the first. To deactivate aliasing, simply alias the font to itself. `\aliasfont` is not recursive. If you type

```
\aliasfont amr10=cmr10 \aliasfont cmr10=mvr10
```

VTEX will perceive the `cmr10` font as `mvr10`; it will, however, perceive `amr10` as `cmr10`.

2.10 Character Set: Letters, Digits, Punctuation

TOPICS ▷ -, –, —, ”, “, ¡, ¿, `\dots`, `/’`, `/”`, `"`, `>]`, `>)`, `[<`, `(<`

A font is usually defined as a *collection of character forms sharing a common design*. This definition is necessarily vague; it does not specify how large the collection should be. In this and the subsequent sections we list all the characters that should or may occur in a font; we will also introduce appropriate TEX commands for accessing these characters.

We will start by stipulating that any English language font should contain 26 letters and 10 digits; the majority of the fonts would actually contain two sets of letters, uppercase and lowercase. Lowercase letters are often omitted in Decorative/Headline fonts:

TROJAN:A B C D E F A B C D E F

ENGRAVER:ABCDEF ABCDEF

✔ If a text font lacks a lower-case letter, VTEX automatically substitutes upper case.

To be of any use, the font should also contain the following standard punctuation characters: period [.], comma [,], colon [:], semicolon [;], left and right parentheses, and brackets [(], [)], [[], []], dash [-], slash [/], question [?] and exclamation [!] marks, single and double quotes. Fonts designed for typesetting would also contain longer and shorter dashes, and usually distinguish between opening, closing, and neutral quotes. The majority of the symbols discussed so far can be directly entered from the keyboard. The dashes and quotes require a bit of extra work.

- **Dashes.**

TEX supports three kinds of dashes: hyphen [-], en-dash [–] and em-dash [—]. They are entered by typing one, two or three dashes.

- **Quotes.**

Use left quotes (‘) once or twice to enter single [‘] or double [“] opening quotes; use right quotes (’) once ([’]) or twice ([”]) to close the quotes.

EXAMPLE 2.35 ▷ Typing

```
Pages 50--71 of ``Ultra-Modern Music Design---Turning
Symphonies Into Nightmares'' explain `cacophonies'.
```

produces

Pages 50–71 of "Ultra-Modern Music Design—Turning Symphonies Into Nightmares" explain 'cacophonies'.

If quotes are nested, as in " 'Murder', she said," separate different sets of quotes with the tie-in character ~ (``~`Murder', she said''. See §3.9).

✔ Neutral quotes are missing in standard TEX. In VTEX, the double quote sign " represents an upright double quote, whereas "" is translated into an upright single quote. VTEX also supports the following additional forms of quotation symbols: German low quotes (‚ and „), entered as /' and /", and single and double guillemets (‹, ›, « and »), entered as (<, >), [< and >]. Under standard TEX, none of these characters is available; the double quote sign " is translated into a double closing quote.

- **Inverted Spanish question and exclamation marks.**

For Spanish typesetting, TEX includes the inverted question ¿ and exclamation ¡ marks. These characters are entered as !` and ?`.

- **Ellipses.**

To ensure proper spacing, ellipses (triple dots) should be entered with the `\dots` command.

EXAMPLE 2.36 ▷ Typing

```
money...␣money\dots\␣money
```

results in

money... money. . . money

2.11 Character Set: Special Characters

TOPICS ▷ +, -, *, {, }, <=, >=, /], \P, \S, \dag, \ddag, +!, +!!, \D

- **Mathematical Characters.**

Most of the fonts also include characters not generally needed for writing text but handy for simple formulas. These characters include the equal sign =, plus +, minus −, asterisk * and braces { and }. The latter are entered as `\{` and `\}`, while =, +, - and * are on the keyboard.

✔ The following additional symbols are supported by VTEX:

- less [<] and greater [>] signs, entered as is
- less or equal [≤] and greater or equal [≥] signs, entered as <= and >=
- plus-minus [±], entered as +-
- divide [÷] entered as /]
- degree [°] entered as \D

In Standard TEX, these characters are available only in math mode (see Chapter 4.)

- dagger [†], double dagger [‡], section [§] and paragraph [¶] signs.

Standard TEX fonts do not include these characters. Instead, they are "borrowed" from mathematical fonts, using the commands shown below:

Symbol	Command	Name
§	\S	Section Sign
¶	\P	Paragraph Sign
†	\dag	Dagger
‡	\ddag	Double Dagger

✔ In VTEX, these characters are included in each font. Alternatively, additional commands `+!` and `+!!` can be used to obtain the daggers [†] and [‡].

EXAMPLE 2.37 ▷ Typing

```
See \S 1 and \S\S\ 4 and 5 for more details.
```

produces

> See §1 and §§ 4 and 5 for more details.

➡ No space is inserted after special symbols. If a space is needed, the special symbol command must be followed by the visible space command (\␣).

- Financial

The dollar sign character [$] appears in most TEX fonts, with the exception of italics, where it is printed as a British Pound. It is entered as \$. To access the Pound sign in standard TEX, the user needs to switch to the italics font and enter \$. In VTEX, the following additional characters are available in all fonts:

- Pound [£], entered as \B
- Cent [¢], entered as \C
- Frank [f], entered as \F
- Yen [¥], entered as \Y

- Other Symbols. The following additional symbols are present in TEX fonts:
  - Ampersand [&], entered as \&
  - At Sign [@], entered as @
  - Per cent [%], entered as \%
  - Per thousand [©], entered as \T, VTEX only.

A large variety of additional symbols is available in specialized mathematical or (VTEX) dingbat fonts. Most of these symbols can only be accessed in math mode (see Chapter 4). They are listed in Appendix 1. These symbols include:

- uppercase and lowercase Greek letters (A, α, B, β, Γ, γ, ..., Ω, ω)
- musical notation ($\flat$, $\sharp$)
- mathematical symbols ($\sum$, $\int$, $\equiv$, ∞)
- card suits ($\heartsuit$, $\spadesuit$, $\diamondsuit$, $\clubsuit$)

To typeset the symbols listed in Appendix 1, you must put the appropriate commands between dollar signs ($...$).

EXAMPLE 2.38 ▷ Typing

```
{\sc Note} $\triangleright$\ The suits $\spadesuit$ and
 $\clubsuit$ are more valuable in some card games.
```

results in

NOTE ▷ The suits ♠ and ♣ are more valuable in some card games.

Special symbols may also be specified by their position number in a font table. Appendix 2 lists several font tables and explains their use.

EXAMPLE 2.39 ▷ The '␣' (space) character is character 32 of the `typewriter` font. It can be accessed by typing `{\tt\char32}`:

```
The symbol '{\tt \char32}' shows that a space is inserted.
```

This will appear as

The symbol '␣' shows that a space is inserted.

You can use the \def command to create short names for commonly used symbols (see §6.1).

2.12 Accents and Foreign Characters

TOPICS ▷ \`, \', \^, \", \ , \=, \., \u, \v, \H, \t, \c, \d, \b, \oe, \OE
▷ \ae, \AE, \aa, \AA, \o, \O, \l, \L, \ss, \i, \j, fi, fl, ffi, ffl, ff

With TEX, you can typeset most languages that use the Latin alphabet; the additional characters and accents are built into most of the fonts. Languages that use other alphabets, such as Russian, Hebrew, or Arabic, require specially designed fonts.

Here are the available accents (both above and under the letter):

Glyph	Command	Name	Appears in
ò	\` o	grave	French, Italian
ó	\' o	acute	French, Italian, Spanish, Polish
ô	\^ o	circumflex	French, Portuguese, Romanian
ö	\" o	umlaut or dieresis	German, French, Swedish
õ	\~ o	tilde	Spanish, Portuguese
ō	\= o	macron	
ȯ	\. o	dot	
ŏ	\u o	breve	Romanian
ǒ	\v o	háček or "check"	Czech, Slovak
ő	\H o	Hungarian umlaut	Hungarian
o͡o	\t oo	tie-after accent	
o̧	\c o	cedilla	French, Portuguese, Romanian
ọ	\d o	dot-under accent	
o̱	\b o	bar-under accent	
ǫ	\g o	ogonek accent	Polish, in VTEX only.

The languages mentioned above do not necessarily form a complete list. The 'o' may be replaced by any other character. TEX will automatically adjust the height of the accent to fit the character which takes it.

The macron, the dot accent, and the tie-after accent are not used in any language *per se*, but, together with the breve accent, they sometimes appear in dictionaries or language texts as pronunciation aids.

TEX also provides a dotless 'ı' and 'ȷ', obtained by typing '\i' and '\j'. They are used when 'i' and 'j' must receive an accent.

Some languages have special letters and ligatures:

Glyph	Command	Name	Appears in
œ,Œ	\oe,\OE		French
æ,Æ	\ae,\AE		Latin, Scand. languages
å,Å	\aa,\AA		Scandinavian languages
ø,Ø	\o,\O		Scandinavian languages
ł,Ł	\l,\L	suppressed-L	Polish
đ,Đ	\x,\X		Croatian, VTEX only
ð,Ð	\eth,\Eth	eth	Old English, VTEX only
Þ,þ	\thorn,\Thorn	thorn	Old English, VTEX only
ß	\ss	es-zet or sharp S	German

EXAMPLE 2.40 ▷ Typing

```
A na\"\i ve fable of \AE sopus recounts that Se\~nor
\OE dipus, going from K\o benhavn to \L\'od\'z during
 the Anschlu\ss, met with the F\"uhrer to
discuss the \ae sthetics of am\oe b\ae.
```

results in

> A naïve fable of Æsopus recounts that Señor Œdipus, going from København to Łòdż during the Anschluß, met with the Führer to discuss the æsthetics of amœbæ.

Notice that the accent command precedes the letter which is to receive the accent.

In most fonts, TEX forms ligatures automatically. The combinations `fi`, `ffi`, `fl`, `ff`, and `ffl` are joined into a single symbol. Thus '`ff`' becomes [ﬀ], '`fi`' becomes [ﬁ], '`fl`' becomes [ﬂ], '`ffl`' becomes [ﬄ], and '`ffi`' becomes [ﬃ].

To prevent the formation of ligatures, use braces to separate the characters which would otherwise be joined.

EXAMPLE 2.41 ▷ Typing

```
Compare ef{f}icient and efficient,
 af{f}luent and affluent.
```

results in

> Compare efficient and eﬃcient, affluent and aﬄuent.

✔ VTEX fonts also recognize Dutch ligatures IJ and ij. These ligatures are automatically formed from 'IJ' and 'ij'.

2.13 Use of Cyrillic Fonts

TEX is not limited to Latin-based alphabets: it can be used just as well to typeset Cyrillic, Greek, and about any other language. In most cases, the issue is not how to make TEX typeset in a foreign language, but rather how to make it more convenient for the program's user. A particular important consideration is that many computers simply cannot display foreign-language characters on the screen, and thus require coding the input in English. In this section we will describe two alternative approaches: one where the foreign letters are coded through the Latin letters, and the other where they are entered directly.

Before getting into a discussion of Cyrillic fonts, it is necessary to mention that there is really no such thing as Cyrillic language or even an alphabet. Historically, Cyrillic writing was invented about 1000 years ago by Greek missionaries Cyrill and Methodius sent to seed the Byzantine culture over the wild plains of what would become Russia. Instead of inventing a totally new alphabet, they took most of the existing Greek and enlarged it with enough new letters to handle the sounds that did not exist in Greek. The original Cyrillic writing went through many revisions over the centuries; the majority of them aimed at reducing the number of different letters. The evolution of the alphabet among different peoples using Cyrillic was different; as result the alphabet used by Russian differs from the one used by Ukrainians or Serbs. Here we say Cyrillic when talking collectively about all of these alphabets; when we talk about Cyrillic fonts we mean fonts large enough to contain all of the possible Cyrillic characters.

An English-language writer that who to incorporate short fragments of Cyrillic text will generally prefer to enter it using English letters. Ideally, one would like to look for a one-to-one correspondence between English and Cyrillic letters; this, however, is not possible since a typical Cyrillic-based alphabet is larger than a Latin-based one. For instance, Russian has 33 letters as opposed to 26 in English; pre-1918 Russian had 36, Ukrainian also includes several borrowed Latin letters that coexist with Cyrillic, while Serbo-Croatian includes special combination letters to depict softer versions of consonants. The correspondence between Cyrillic and English is usually established by identifying similar letters in both alphabets, mapping extra Latin letters in a somewhat arbitrary fashion, and handling the remaining Cyrillic letters through ligatures or macros. For example, the manuscript may contain English letters '`A`', '`Z`', and '`Q`'; when a Cyrillic font is selected by TEX, these letters will become А, З and Ч. Notice that in the case of '`A`', the correspondence is obvious; for '`Z`' it is phonetic (both English '`Z`' and Cyrillic 'З' denote the same sound; for '`Q`' it is totally arbitrary: the letter '`Q`' actually represents the sound '`ch`'. According to the standard developed in the University of Washington, other Cyrillic letters are handled as phonetic-based ligatures. Thus, typing 'ya' results in я, and typing '`sh`' results in ш. Some of the ligatures are not that intuitive. For example, the Russian alphabet includes symbols ь (soft sign) and ъ (hard sign). These symbols do not produce any sound of their own; instead they either soften or harden the sound of the preceding consonant. According to the University of Washington convention, these symbols are entered as '`p1`' and '`p2`'— punctuation signs 1 and 2. Furthermore, while Russian makes letters softer by appending a soft sign, Serbo-Croatian builds ligatures of its own. For example, soft 'N' and 'L' in these two languages will appear as Нь and Ль (enter '`Np1`' and '`Lp1`'), and, correspondingly as Њ and Љ (enter '`N1`' and '`L1`'). The University of Washington convention includes a total of 78 ligatures ranging from very intuitive to not-at-all obvious.

The ligature scheme has a few serious drawbacks. One is that for an input text longer than just a few words, reading or writing it becomes an exercise in cryptology. A second is that sometimes ligatures have undesirable side effects. For instance, many Russian words allow the sound '`s`' to be followed by the sound '`h`'. The word for scheme is схема (skhe-ma); the ligature mechanism would corrupt it into шема (she-ma). While a simple way to prevent a ligature formation is to insert a couple of extra braces (enter `s{h}ema` or `s{}hema`, not `shema`), this does not simplify things for a user.

A different approach is to combine TeX with a software utility that remaps unneeded high-ASCII characters into Cyrillic letters. After a coding scheme is selected, one needs to provide three separate but compatible components:

- A graphics screen driver that will display Cyrillic letters on the terminal
- A keyboard driver that will allow easy entering of these letters
- A TeX translation utility that will substitute correct characters out of TeX fonts.

Many terminal and keyboard drivers were developed that allow entering of the Cyrillic characters. These drivers differ from country to country; the layout of screen fonts in the former Soviet Union while identical to the one used in Bulgaria is quite different from the one used in Yugoslavia. The translation done by TeX is similar in all cases, except that different tables may be used depending on the layout of screen and TeX fonts.

For example, in the so-called Alternative Coding Scheme, widely accepted in the former Soviet Union, Cyrillic л corresponds to the ASCII character 140. In a TeX font file this character can be present either under the same number, or, more likely, to be in the position of the English '`l`' (`\char108`). To force TeX to use `\char108` when `\char140` actually appears in the input stream, one would use the following commands:

```
\catcode140 = \active \def'\140{\char108}
```

(`\catcode` and `\def` are explained further on in this book.)

Generally, this approach is substantially easier for the user when large portions of text need to be typeset.

One additional consideration should be given to the option of combining Cyrillic and Latin fonts in one font. Often, Cyrillic fonts are designed based on their Latin counterpart, but packaged as separate fonts. The user has to explicitly switch between Cyrillic and Latin via font commands. Another disadvantage is that in most cases such fonts are usually designed by different people, which results in the design of Cyrillic letters being inferior to that of the Latin. Alternatively, one may design a font that combines both Cyrillic and Latin letters. This would be preferable for typesetting text that often switches between Latin and Cyrillic. Ideally, such fonts would follow one of the accepted layouts, such as the Alternative Coding Scheme, to eliminate the need for TeX translation. In all cases, such fonts should distinguish between even identical letters appearing in both alphabets; thus Latin `A` should be a different symbol from the Cyrillic `A`, even if the two look completely alike. This is necessary, in particular, to allow two different hyphenation tables to coexist.

We conclude this section with some samples of Cyrillic fonts. Below is the appearance of Computer Modern Cyrillic (designed by the University of Washington) followed by MicroPress's Cyrillic Helv. and Balt.

Кириллица: А Б В Г Д Е а б в г д е

Кириллица: А Б В Г Д Е а б в г д е

Кириллица: А Б В Г Д Е а б в г д е

Chapter 3

Formatting

3.1 Page Layout

TOPICS ▷ `\hsize`, `\vsize`, `\hoffset`, `\voffset`, `\landscape`

Paper manufacturers measure only the physical size of the paper. Typesetters are also concerned with the size of the *printable page*, because printing is almost never done on the entire physical page. Margins are usually left blank; in computerized typesetting margins must frequently be left blank because of hardware limitations. For example, most current laser printers view $\frac{1}{4}$-inch margins as "dead space"—nothing can be printed there.

The figure below shows the layout of a page:

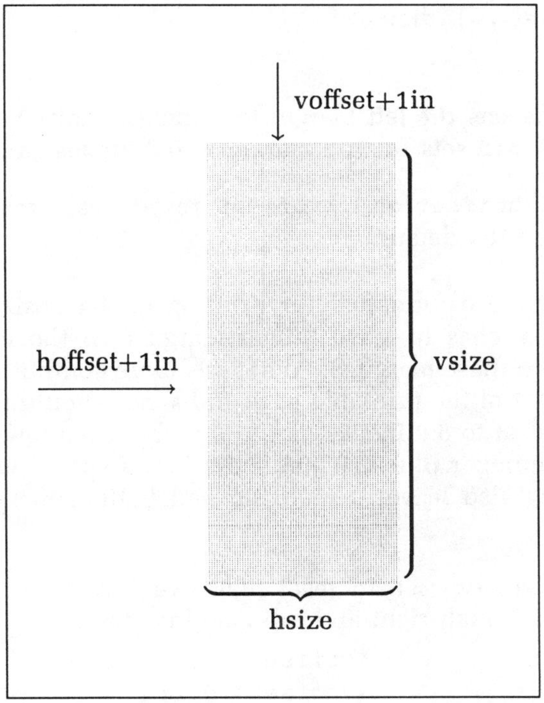

The gray area in the figure represents the printable page. Its dimensions are specified by the `\hsize` (horizontal size) and `\vsize` (vertical size) commands.

SYNTAX ▷ `\hsize` [=] ⟨dimen⟩
▷ `\vsize` [=] ⟨dimen⟩

where ⟨dimen⟩ specifies the size in inches (`in`), printer points (`pp`), centimeters (`cm`), millimeters (`mm`), picas (`pc`), and a few other units (see §5.3 for the complete list).

EXAMPLE 3.1 ▷ To make the printable page 3 inches wide and 5 inches high, enter

```
\hsize=3in  \vsize=5in
```

or, without the optional equal sign,

```
\hsize 3in  \vsize 5in
```

The position of the printable page on the physical page is determined by the '`\hoffset`' and '`\voffset`' commands. If neither of these commands is used, the left and top margins are set by default to 1 inch, which is the standard for most documents. If you specify '`\hoffset`', the left margin becomes 1 inch plus the value of '`\hoffset`'. Similarly, the top margin is 1 inch plus the value of '`\voffset`'.

SYNTAX ▷ `\hoffset` [=] ⟨dimen⟩
▷ `\voffset` [=] ⟨dimen⟩

EXAMPLE 3.2 ▷

`\hoffset=1in` sets the left margin to 2 inches, since $1+1=2$.
`\voffset=-0.3in` sets the top margin to 0.7 inches, since $1.0+(-0.3)=0.7$.

Negative values for `\hoffset` and `\voffset` respectively make the left and top margins narrower than the default.

By default, documents are formatted for printing on letter-size paper (8.5"×11"). `\hsize` is set to 6.5 inches to leave 1-inch margins on the left and on the right (`\hoffset` may change these margins). `\vsize` is more complicated because headers and footers are not part of the printable page and space should be reserved for them. By default, `\vsize` is set to 8.8 inches, not to 9 inches, to allow for a 0.1-inch footer containing the page number plus 0.1 inch extra "dead space" needed by some laser printers. This setting also leaves 1-inch top and bottom margins (`\voffset` may change these margins).

EXAMPLE 3.3 ▷ To typeset on legal (8.5"×14") paper with 1.5 inch left and bottom margins and 0.5 inch right and top margins, enter

```
\hsize=6.5in          \hoffset=0.5in
\vsize=11.9in         \voffset=-0.5in
```

CAUTION ▷ The page configuration parameters should be set only at the beginning of a document, or immediately after a page break (see §3.14), and should not be changed in the middle of a page.

To typeset in landscape mode, the user needs to swap the values of `\hsize` and `\vsize`. VTeX's `\landscape` command does exactly this (see §11.9).

➡ The settings of `\hoffset` and `\voffset` may be overridden by the printer driver margin options (check your driver manual); `\vsize` and `\hsize` cannot be overridden since they define the actual size of the printable page. The right and bottom margins are not explicitly given since they are automatically determined from the values of the four parameters mentioned above and the physical size of the paper.

Most typesetting is done in an `\hsize` by `\vsize` rectangle. TeX can access two more areas: the area just below the printable page is reserved for footers and the area just above the printable page may be used for headers. By default, TeX produces no headers; the footer line contains a centered page number. (§3.3 describes the commands used to format these two areas.)

In general, the printable page is composed of lines and paragraphs separated by vertical spacing. Formatting a page consists of the following steps:

- *Line formatting:* centering, right- and left-justification. This includes formatting headers and footers, since they are just special cases of lines (see §3.2,§3.3).
- *Paragraph formatting:* indentation, paragraph shaping, line breaks, and hyphenation (see §3.6,§3.7).
- *Page formatting:* spacing between lines and paragraphs, page breaks (see §3.13,§3.14).

3.2 Lines of Text

TOPICS ▷ `\line`, `\centerline`, `\leftline`, `\rightline`

A line of text can be formatted in four basic ways:

- `\centerline` centers the text
- `\leftline` left-justifies the text
- `\rightline` right-justifies the text
- `\line` spaces words evenly between the margins

All four commands have the same syntax:

SYNTAX ▷ `\centerline` ⟨argument⟩
▷ `\leftline` ⟨argument⟩
▷ `\rightline` ⟨argument⟩
▷ `\line` ⟨argument⟩

where ⟨argument⟩ is the text to be formatted.

EXAMPLE 3.4 ▷ Typing

```
\line{Mary had a little lamb.}
\leftline{Mary had a little lamb.}
\centerline{Mary had a little lamb.}
\rightline{Mary had a little lamb.}
```

results in

Mary	had	a	little	lamb.
Mary had a little lamb.				
		Mary had a little lamb.		
				Mary had a little lamb.

➡ The `\line` command often causes excessive spacing between words. If this happens, you will get a message similar to

```
Underfull \hbox (badness 10000) detected at line 199
\tenrm Mary had a little lamb.
```

warning you that the current line (*horizontal box*, according to TeX) is badly formatted (see §3.11, §8.3).

➡ Horizontal spacing and glue commands, introduced in §3.4 and §5.4, may be used for constructing more complex formats.

3.3 Headers and Footers. Page Numbers

TOPICS ▷ `\headline`, `\footline`, `\pageno`
▷ `\nopagenumbers`, `\advancepageno`, `\folio`

When TeX formats a page, it reserves some space just above and below the body of the text for two special lines: the header and the footer.

The `\headline` and `\footline` commands are used to format these lines. These commands are so named in order to emphasize that the header and the footer each occupy only one line.

SYNTAX ▷ `\headline [=] {` ⟨headline description⟩ `}`

where ⟨headline description⟩ contains formatting instructions and the text to be placed in the header. Similarly,

SYNTAX ▷ `\footline [=] {` ⟨footline description⟩ `}`

defines the format of the page footer. In both cases the equal sign is optional but the braces must be typed in.

EXAMPLE 3.5 ▷ After the command

```
\headline={\centerline{\rm Joe's confidential report}}
```

is issued, all pages will contain the centered headline:

Joe's confidential report

set in the `\rm` font.

It is no accident that the font is explicitly specified. If it were not, each headline would be typeset in the font which is current at the beginning of the page the headline is on. Thus, some headlines might appear in roman font, some in italics, and some might be illegible, being set in a symbol font.

Headers or footers usually contain a page number, which is controlled by the `\pageno` and `\folio` commands.

The `\pageno` command sets the current page number.

SYNTAX ▷ `\pageno` [=] ⟨page number⟩

where ⟨page number⟩ is an integer number.

EXAMPLE 3.6 ▷ By default, page numbering starts at 1. To start with a different page number (3, for instance), specify

```
\pageno=3
```

at the beginning of the document.

➡ If a complete document can not be processed at once, perhaps because of a lack of disk space, the document may be split into smaller files. Make sure that each file starts on a new page (see §3.14); `\pageno` should be set correctly in order to assure continuous page numbering. For example, if the first file of the document produces pages 1 through 124, start the second file with `\pageno=125`.

The `\folio` command inserts the current page number in the text.

EXAMPLE 3.7 ▷ Typing

```
You are now reading page \folio.
```

produces

You are now reading page 61.

This command is most often used inside headers and footers.

EXAMPLE 3.8 ▷ Specifying

```
\headline={\tenrm\centerline{- \folio\ -}}
```

will generate a headline similar to

— 62 —

on each page. Instead of 62, each headline will, of course, contain the actual page number.

EXAMPLE 3.9 ▷ By default, the headline is left blank and the footline contains a centered page number. To reverse the formatting of the headline and the footline, type

```
\footline={}
\headline={\tenrm\centerline{\folio}}
```

The `\advancepageno` command is used to advance the page count by one. This command is automatically executed each time a new page is started. You may use it to skip a page in order to provide space for a picture or a diagram that is not printed by TeX.

SYNTAX ▷ `\advancepageno`

➡ This command does not cause a page break.

`\pageno` may be given a positive or a negative value. Negative page numbers do not normally occur in books; instead, `\folio` typesets them as the corresponding positive roman numerals.

EXAMPLE 3.10 ▷ Typing

```
\pageno = -34
\headline={\tenrm\centerline{- \folio\ -}}
```

will print the headline

— xxxiv —

on the next page.

For consistency, `\advancepageno` increases the value of `\pageno` by one if it is positive, and decreases it by one if it is negative. Thus, page "-1" (i) is followed by page "-2" (ii), then by page "-3" (iii), and so on.

Many books have two separate numbering sequences: one for the beginning pages (table of contents, introduction, etc.) and another for the main body of the book. The first sequence is usually in roman numerals (i, ii, iii, iv, ..., xii, for example) and the second is in arabic numerals (1, 2, 3, ...).

To do this, type `\pageno=-1` at the beginning of the document, then type `\pageno=1` at the beginning of the main body of the document.

The command `\nopagenumbers` suppresses printing of the page numbers. It should be used when typesetting letters and short memoranda.

SYNTAX ▷ `\nopagenumbers`

Odd and even pages may contain differently formatted headers and footers (see §6.9).

3.4 Default Horizontal Spacing

TOPICS ▷ `\frenchspacing`, `\nonfrenchspacing`, `\null`, `\␣`

TEX provides three types of interword spacing:

- ordinary: usually between words not separated by punctuation marks
- between parts of a sentence: usually after commas, colons and semicolons
- between sentences: usually after periods, exclamation marks and question marks

In general, the ordinary space is the smallest, the space between parts of a sentence is a bit larger, and the space between sentences is the largest.

EXAMPLE 3.11 ▷ Typing

```
\line { "Oh, Galaxy!," cursed the small green man.
His ship was ruined.}
```

produces

"Oh, Galaxy!," cursed the small green man. His ship was ruined.

The difference between the three types of spacing is more visible if the text is typeset sparsely. Here the `\line` command was used to this end.

To use the correct spacing, TEX must first find the ends of sentences. Notice that not every period (or exclamation point or question mark) ends a sentence. For example, in

"Error!!!," A. B. Jones exclaimed.

it would be wrong to insert extra space after the exclamation points or after the first two periods. TEX uses the following rule: a sentence ends only in a period (exclamation point, question mark) that does not come after a capital letter and is followed by a space. The example above, as well as most sentences, will be typeset correctly.

However, this rule does not always work. Space will be missing after a sentence ending with a capital letter. At the same time, an abbreviation inside a sentence will be followed by an unnecessary space. The next example shows how to circumvent this rule.

EXAMPLE 3.12 ▷ Typing

```
\line{Consult the Rep. of Congr. Comm. on New Appl. of
High Tech. on this matter.}
\line{Jones works for the FDA. Brown works for the State.
Smith works for himself.}
```

results in

Consult the Rep. of Congr. Comm. on New Appl. of High Tech. on this matter. Jones works for the FDA. Brown works for the State. Smith works for himself.

To suppress the extra space after abbreviations, insert the \␣ command immediately after the period. The first line of the above example should be entered as

```
\line{Consult the Rep.\␣of Congr.\␣Comm.\␣on New Appl.\␣
of High Tech.\␣on this matter.}
```

to get

Consult the Rep. of Congr. Comm. on New Appl. of High Tech. on this matter.

This works because the period is no longer followed by a space but by a command that generates the ordinary interword spacing.

➡ A line should not break after certain abbreviations, including initials. To prevent that, use the ~ (tilde) instead of \␣. For example, type `Mr.~C.~Jones` instead of `Mr.\␣C.\␣Jones` (see §3.9).

In the second line of the above example, the trick is to separate the capital letter from the period. The `\null` command is used for this purpose; the example should be entered as

```
\line{Jones works for the FDA\null.
Brown works for the State.
Smith works for himself.}
```

to get

Jones works for the FDA. Brown works for the State. Smith works for himself.

The `\frenchspacing` command makes all three types of spaces equal. `\nonfrenchspacing` restores the default.

SYNTAX ▷ `\frenchspacing`
▷ `\nonfrenchspacing`

EXAMPLE 3.13 ▷ The following sentence is first typeset in the usual way and then with `\frenchspacing`. Typing

```
\line{"Veni, vidi, vici," said Mr.\ Caesar to the
    press. He knew he would be quoted.}
{\frenchspacing
\line{"Veni, vidi, vici," said Mr.\ Caesar to the
    press. He knew he would be quoted.}}
```

produces

"Veni, vidi, vici," said Mr. Caesar to the press. He knew he would be quoted. "Veni, vidi, vici," said Mr. Caesar to the press. He knew he would be quoted.

`\frenchspacing` should be placed inside a group (braces) in order to prevent it from changing the spacing in the rest of the document.

3.5 Horizontal Skips

TOPICS ▷ `\hskip`, `\enspace`, `\quad`, `\qquad`, `\␣`, `\indent`

Horizontal skips are used to override the default spacing by specifying the desired amount of space in either absolute or font-dependent units.

The `\hskip` command is used to specify the amount of horizontal spacing in absolute units (inches, centimeters, points, etc.—see §5.3 for the list).

SYNTAX ▷ `\hskip` ⟨dimen⟩

EXAMPLE 3.14 ▷ Typing

```
A\hskip.1inB\hskip.2inC\hskip.3inD\hskip.4inE\hskip.5inF
```

results in

A B C D E F

notice that the space after 'in' is unnecessary.

`\␣`, `\enspace`, `\quad` and `\qquad` are horizontal skips by predefined amounts that depend on the current font. For each font, TEX remembers the amount of normal interword space.

- `\␣` (backslash followed by a space) produces a space as wide as the normal interword space in the current font
- `\enspace` produces space as wide as the en-dash '–' of the current font
- `\quad` is equivalent to two `\enspace`'s
- `\qquad` is equivalent to two `\quad`'s

EXAMPLE 3.15 ▷ Typing

```
\rm a\ b\enspace c\quad d\qquad e

\tt a\ b\enspace c\quad d\qquad e
```

results in

a b c d e `a b c d e`

EXAMPLE 3.16 ▷ The NOTES in this book were typeset with

```
{\sc Note} $\triangleright$ \qquad
 This is the text of the note.
```

which gives

NOTE ▷ This is the text of the note.

The `\indent` command produces a horizontal skip as wide as the paragraph indentation (see §3.6).

3.6 Paragraph Formatting

TOPICS ▷ `\par`, `\indent`, `\noindent`, `\parindent`, `\raggedright`

The simplest way to end a paragraph is to leave an empty line after it. Alternatively, the `\par` command produces the same effect as an empty line.

SYNTAX ▷ `\par`

EXAMPLE 3.17 ▷ Typing

```
It is easy to describe a grue, but hard to imagine one.

Think of a twenty-foot long beast, covered with both scales
and long hair.  Two pairs of small green eyes are hidden
between three pairs of horns.  Imagine also five rows of
teeth and three long tails. \par Do not be overtly
concerned if you cannot imagine a grue--nobody has ever
seen one.
```

produces three paragraphs:

> It is easy to describe a grue, but hard to imagine one.
>
> Think of a twenty-foot long beast, covered with both scales and long hair. Two pairs of small green eyes are hidden between three pairs of horns. Imagine also five rows of teeth and three long tails.
>
> Do not be overtly concerned if you cannot imagine a grue—nobody has ever seen one.

By default, paragraphs are right-justified. The `\raggedright` command suppresses justification, making the right margin uneven. The interword spacing in a paragraph formatted with `\raggedright` is uniform.

SYNTAX ▷ `\raggedright`

EXAMPLE 3.18 ▷ Typing

```
\raggedright
By the late twenty-seventh century, the successes of small
off-planet manufacturers forced such giants as
Interplanetary Machines for Business (IMB) to move part of
their own production off-planet.  A substantial
restructuring of the company became necessary and millions
of workers were laid off.  In a desperate attempt to hold
on to the market, IMB introduced yet another incompatible
series of products.
```

produces

> By the late twenty-seventh century, the successes of small off-planet manufacturers forced such giants as Interplanetary Machines for Business (IMB) to move part of their own production off-planet. A substantial restructuring of the company became necessary and millions of workers were laid off. In a desperate attempt to hold on to the market, IMB introduced yet another incompatible series of products.

➡ The `\raggedright` command affects the rest of the document. If only a few paragraphs need be ragged, they should be placed inside a group.

The `\parindent` command sets the amount of paragraph indentation. The default value is 20 points (about 0.28 inches).

SYNTAX ▷ `\parindent` ⟨dimen⟩

where ⟨dimen⟩ is the paragraph indentation. For example,

```
\parindent 2cm
```

sets paragraph indentation to two centimeters. The command

```
\parindent 0pt
```

suppresses paragraph indentation by setting it to 0.

The `\noindent` command is used at the beginning of a paragraph to suppress indentation, even if `\parindent` is not set to zero.

SYNTAX ▷ `\noindent`

The `\indent` command skips a horizontal space equal to the current paragraph indentation (`\parindent`).

SYNTAX ▷ `\indent`

➡ Neither of the last two commands forces a new paragraph.

EXAMPLE 3.19 ▷ Typing

```
Gertrude Stein said

\centerline{A rose is a rose is a rose is a rose.}

\noindent which is frequently misquoted as
```

```
\centerline{A rose is a rose is a rose.}
```

results in

> Gertrude Stein said
>
> A rose is a rose is a rose.
>
> which is frequently misquoted as
>
> A rose is a rose is a rose.

The `\noindent` command is also used after diagrams, tables, or equations inserted into the middle of a sentence.

➡ Empty lines around `\centerline` are redundant; the same holds for the other line commands (`\line`, `\leftline`, and `\rightline`).

3.7 Non-Standard Paragraph Formatting

TOPICS ▷ `\hangindent`, `\textindent`, `\hangafter`, `\leftskip`, `\rightskip`
▷ `\narrower`, `\parshape`

Paragraphs may be formatted in more ways than those described in the previous section.

Some paragraphs are indented on more than one line. This is called *hanging indentation*. To define hanging indentation, use both the `\hangindent` command to set the size of the indentation and the `\hangafter` command to specify which lines are to be indented.

SYNTAX ▷ `\hangindent` [=] ⟨dimen⟩

where ⟨dimen⟩ is the hanging paragraph indentation. It can be either positive or negative—in the latter case the indentation is done from the right.

SYNTAX ▷ `\hangafter` [=] ⟨number⟩

where ⟨number⟩ is the number of lines to be indented. It can be either positive or negative. If ⟨number⟩ is positive, all lines starting with the line ⟨number⟩ are indented. If ⟨number⟩ is negative, the first −⟨number⟩ lines are indented.

EXAMPLE 3.20 ▷ Typing

```
\hangindent = 1in \hangafter = -3 \parindent=0pt
The Royal House of the Banana Islands Sultanate decreed
\TeX\ to be the official typesetting software of this
state.  A later decree mandated that the first three lines
of any official document shall be indented exactly one
```

```
inch in order to leave space for the Royal Seal.  Law
enforcement officials were instructed to view failure to
comply as treason.
```

results in

> The Royal House of the Banana Islands Sultanate decreed TEX to be the official typesetting software of this state. A later decree mandated that the first three lines of any official document shall be indented exactly one inch in order to leave space for the Royal Seal. Law enforcement officials were instructed to view failure to comply as treason.

The usual paragraph indentation (`\parindent`) works independently of the hanging indentation. It was suppressed to avoid double-indenting the first line of the example.

By default, `\hangindent` is set to 0 (no hanging indentation) and `\hangafter` is set to 1 (all lines are indented).

The `\leftskip`, `\rightskip`, and `\narrower` commands indent the entire paragraph (all lines) on the left, right and both, respectively.

SYNTAX ▷ `\leftskip` ⟨dimen⟩

indents the entire paragraph on the left by ⟨dimen⟩.

SYNTAX ▷ `\rightskip` ⟨dimen⟩

indents the entire paragraph on the right by ⟨dimen⟩.

SYNTAX ▷ `\narrower`

indents the entire paragraph both on the left and on the right by the current setting of `\parindent`.

EXAMPLE 3.21 ▷ Typing

```
The Constitution of the United States of America
begins with the following lines:

\leftskip 1.0in \rightskip 1.5in
We, the people of the United States, in order to form a
more perfect Union, establish justice, ensure domestic
tranquility, provide for the common defense, promote
the general welfare, and secure the blessings of liberty
to ourselves and our posterity, do ordain and establish
this Constitution for the United States of America.
```

results in

> The Constitution of the United States of America begins with the following lines:
> We, the people of the United States, in order to form a more perfect Union, establish justice, ensure domestic tranquility, provide for the common defense, promote the general welfare, and secure the blessings of liberty to ourselves and our posterity, do ordain and establish this Constitution for the United States of America.

`\narrower` is frequently used to insert large quotations into the text.

The most powerful paragraph formatting command is `\parshape`. It sets the length and the indentation of each line separately.

SYNTAX ▷ `\parshape [=]` $n \quad i_1\ l_1 \quad i_2\ l_2 \quad \ldots \quad i_n\ l_n$

where n is the number of lines for which the length and the indentation are specified. i_k is the indentation and l_k is the length of the k-th line of the paragraph.

EXAMPLE 3.22 ▷ Typing

```
\parshape=5  .1in 3.5in  .3in 3.1in  .5in 2.7in
             .7in 2.3in  .9in 1.9in
\noindent

Some physicists believe that the Universe is constantly
contracting.  Distances are becoming shorter, celestial
bodies are shrinking, gas mileage is getting better,
floppy disks are growing smaller, and the dollar is
falling.

\parshape 0
```

produces

> Some physicists believe that the Universe is constantly
> contracting. Distances are becoming shorter, celes-
> tial bodies are shrinking, gas mileage is get-
> ting better, floppy disks are growing
> smaller, and the dollar is falling.

The first argument of the first `\parshape` command (5) is followed by 5 pairs of numbers that define the dimensions of the first 5 lines of the paragraph. If more than

10 numbers had been given, the extra ones would have been processed as ordinary text and not as part of the \parshape command (they would have appeared in the output). On the other hand, if fewer than 10 numbers had been given, you would have received an error message. In the example, the typeset paragraph contains exactly 5 lines. If there had been more than 5 lines, the ones after the fifth would have been formatted like the fifth line. (No error message is printed.)

\parshape 0 restores the usual form of the paragraphs. The empty line just before \parshape 0 is needed to prevent this command from affecting the paragraph before it. The \parshape in effect at the end of a paragraph is used to format that paragraph.

The \textindent command places some text inside the indentation. It is used to tag a paragraph.

SYNTAX ▷ \textindent ⟨argument⟩ ⟨paragraph text⟩

where ⟨argument⟩ contains the tag. The paragraph to be tagged should follow immediately.

EXAMPLE 3.23 ▷ Typing

```
\textindent{\bf 6} I opened to my beloved; but my beloved
had withdrawn himself, and was gone: my soul failed when
he spake: I sought him, but I could not find him;
I called him, but he gave me no answer.
```

results in

> **6** I opened to my beloved; but my beloved had withdrawn himself, and was gone: my soul failed when he spake: I sought him, but I could not find him; I called him, but he gave me no answer.

3.8 Lists

TOPICS ▷ \list, \enum, \enumenum, \item, \itemitem

TEX has special commands for formatting lists.

The \list command is used to typeset unnumbered lists. It formats each item as a separate paragraph. The first line of each paragraph has a bullet symbol (•) in the indentation field and every line is indented by the value of \parindent.

\list, \enum and \enumenum are not standard (see §11.9).

SYNTAX ▷ \list ⟨item⟩

EXAMPLE 3.24 ▷ Typing

```
Aunt Polly's shopping list:
\list Apples
\list Soap  \list Kiwi fruit
\list Christmas cards for Tom, Dick, Harry, Jack, Jill,
Grumpy, Sleepy, Dopey, and Mrs. Rumpelstiltskin (although
she doesn't deserve one).
```

produces

Aunt Polly's shopping list:

- Apples
- Soap
- Kiwi fruit
- Christmas cards for Tom, Dick, Harry, Jack, Jill, Grumpy, Sleepy, Dopey, and Mrs. Rumpelstiltskin (although she doesn't deserve one).

The \enum command works the same way as \list, except that it replaces bullets with numbers.

SYNTAX ▷ \enum ⟨item⟩

The \enumcount command is used to set the number of the first item on the list (it need not be 1).

SYNTAX ▷ \enumcount [=] ⟨number⟩

The \enumenum command is used to make a list within a list. Such a list is double-indented and tagged with the letters *a, b,...z* (at most 26 items).

SYNTAX ▷ \enumenum ⟨sub-item⟩

EXAMPLE 3.25 ▷ Typing

```
\enumcount = 1
{\bf Agenda of the Board Meeting:}
\enum Election of new corporate officials
\enum Reports of regional offices
\enumenum Brown for the West Coast
\enumenum Smith for the Midwest
\enumenum Jones for the East Coast
```

```
\enum Retirement party for President Bob Pumpernickel III
```

produces

> **Agenda of the Board Meeting:**
> 1. Election of new corporate officials
> 2. Reports of regional offices
>   a) Brown for the West Coast
>   b) Smith for the Midwest
>   c) Jones for the East Coast
> 3. Retirement party for President Bob Pumpernickel III

Primary lists must be initialized with the `\enumcount` command. Secondary lists do not need initialization, but `\enumenum` must appear after `\enum`, because the latter command does the initialization.

The `\item` command is identical to `\enum` or `\list` except that any text may be used instead of bullets or numbers. The ⟨label⟩ is right-aligned in the indentation field.

SYNTAX ▷ `\item {`⟨label⟩`}` ⟨item⟩

EXAMPLE 3.26 ▷ Typing

```
Things to do today: {\parindent 1in
\item{11:00 a.m.} Wake up
\item{11:30 a.m.} Have a bite
\item{12:00 p.m.} Read {\sf THE DAILY GOSSIP}
\item{12:30 p.m.} Take a stroll
\item{ 1:00 p.m.} Have another bite }
```

results in

> Things to do today:
> 11:00 a.m. Wake up
> 11:30 a.m. Have a bite
> 12:00 p.m. Read THE DAILY GOSSIP
> 12:30 p.m. Take a stroll
> 1:00 p.m. Have another bite

The indentation was increased to accommodate the long tags, which do not fit into a standard 20-point (about 0.28 inches) indentation. Tags that do not fit into the indentation extend into the left margin.

The `\itemitem` command produces secondary lists. It is similar to `\enumenum`, except that each item's label is explicitly given.

SYNTAX ▷ `\itemitem {`⟨label⟩`}` ⟨item⟩

3.9 Line Breaks

Topics ▹ ~, \slash, \hbox

TEX formats the text to make the interword spacing as uniform and as close to the ideal as possible. For a human, however, the uniformity of spacing is not the only consideration. If you typed

```
To understand the  psychology of  grues, read the earlier
works of A. B. Jones.  Pay special attention to pp. 34--43
and fig. 12 in Appendix Q.
```

you may get a line break between the initials of the author, after `pp.` or `fig.` or between `Appendix` and `Q`, which usually looks wrong. To avoid this, use a '~' (tilde or *tie-in*) to denote an unbreakable or hard space.

Instead, this example should be typed in as

```
To understand the  psychology of  grues, read the earlier
works of A.~B.~Jones.  Pay  special attention to
pp.~34--43 and fig.~12 in Appendix~Q.
```

to ensure that a line break does not occur in the wrong place.

➡ Using the tie-in also ensures that the space it marks will not be too large.

The tie-in should be used

- between a person's initials: `A.~B.~Jones`, `U~Thant`
- before a numbered reference, as in `page~7`, `fig.~12`
- before certain mathematical symbols: `function~$f$`, `number~13`

TEX tries to break lines on hyphens or dashes, if there are any. If this is undesirable, as in "pp. 34–43" or "page X-2", use the `\hbox` command: `pp.~\hbox{34--43}`, `page~\hbox{X-2}`. The argument of `\hbox` is never split across lines.

Sometimes line breaks should be allowed after the slash '/'. The `\slash` command generates a slash with an optional break. For example, type `read\slash write` instead of `read/write` to allow a line break after `read/`.

3.10 Hyphenation

TOPICS ▷ `\hyphenation`, `\showhyphens`, `\discretionary`
▷ `\lefthyphenmin`, `\righthyphenmin`

Generally, a document will look better if it is typeset with as few hyphenated words as possible. TEX first tries to format a paragraph without hyphenation. If this does not work (the space between words becomes too large or too small), it hyphenates some words and then reformats.

Hyphenation is a difficult problem for computers. Most packages that do it use a large hyphenation dictionary. This requires a lot of disk space, and still does not always ensure correct results. For example, the word "record" should be hyphenated as "rec-ord" if it is a noun and as "re-cord" if it is a verb – and no computer program written to date (including TEX) can always distinguish between nouns and verbs. In addition, there are many proper nouns and foreign words which cannot all be included in hyphenation dictionaries.

Instead of using a large dictionary, TEX remembers the hyphenation rules of English and supplements them with a small table of exceptions. This approach saves storage and ensures correct hyphenation for the large majority of English words. An English word will never be hyphenated in the wrong place. However, TEX might not find all the possible ways to break a word. Words which may not be hyphenated correctly include:

- foreign words and proper nouns
- words hyphenated differently depending on their grammatical function
- less commonly used words that are not included in the exception table

You may use the `\hyphenation` command to modify the table of exceptions to ensure correct hyphenation.

SYNTAX ▷ `\hyphenation{` ⟨word list⟩ `}`,

where ⟨word list⟩ is a list of hyphenated words; only letters, dashes, and spaces may occur in it.

EXAMPLE 3.27 ▷ Typing

```
\hyphenation{re-cord Ar-gen-ti-na}
```

instructs TEX to hyphenate "record" as a verb (by default, it would not hyphenate this word at all in order not to make an error), and gives the complete hyphenation for a foreign word not known to TEX.

The `\showhyphens` command is used to see how a particular word is hyphenated.

SYNTAX ▷ `\showhyphens{` ⟨word⟩ `}`

This command should be used in interactive mode. It is always accompanied by an `underfull` error message—this message does not indicate an actual error and should be ignored (see §3.11).

EXAMPLE 3.28 ▷ Typing (at the ∗-prompt)

```
\showhyphens{Argentina}
```

causes TEX to display

```
Underfull \hbox (badness 10000) detected at line 0
[] \tenrm Ar-gen-ti-na
```

(See also the previous example.)

Some foreign (mostly German) words change their spelling when hyphenated. For example, "Becker" is hyphenated as "Bek-ker", not as "Bec-ker". Use the `\discretionary` command to give the hyphenation of such words.

SYNTAX ▷ `\discretionary{`⟨pre-break⟩`}{`⟨post-break⟩`}{`⟨no-break⟩`}`

where ⟨pre-break⟩ is the part of the word before a hyphen, ⟨post-break⟩ is the part of the word after the hyphen, and ⟨no-break⟩ is the unhyphenated word.

EXAMPLE 3.29 ▷ Enter

```
Mr.~\discretionary{Bek-}{ker}{Becker} lost the tournament.
```

or

```
Mr.~Be\discretionary{k-}{k}{ck}er lost the tournament.
```

instead of

```
Mr.~Becker lost the tournament.
```

to ensure correct hyphenation.

➡ The macro facility of TEX (see §6.1) simplifies entering words like "Becker"(see also §6.4).

In English, when a compound like "self-sufficient" or "willy-nilly" is split across lines at the hyphen, the hyphen appears only at the end of the upper line, just as if the compound were a single word ("selfsufficient" or "willynilly"). This ambiguity is not acceptable in a dictionary or in a grammar book. There a second hyphen is printed at the beginning of the next line. Some languages do this in all cases.

EXAMPLE 3.30 ▷ In such a case, the word "self-sufficient" should be entered as

```
self\discretionary{-}{-}{-}sufficient
```

to ensure that hyphens appear on both lines. As before, the macro facility can be used to simplify entering such text (see §6.1).

Starting with the version 3.0, TEX includes commands that control how many letters may be split off before and after a hyphenation break. These commands are `\lefthyphenmin` and `\righthyphenmin`.

SYNTAX ▷ `\lefthyphenmin`⟨number⟩
▷ `\righthyphenmin`⟨number⟩

where the defaults are 2 and 3. This effectively determines that the shortest word that can be hyphenated is at least five-letters long; the only way to hyphenate a five letter word is after the second letter.

3.11 Overfull and Underfull Lines

TOPICS ▷ `\tolerance`, `\hbadness`, `\vbadness`, `\hfuzz`, `\vfuzz`, badness

As explained in the previous section, TEX tries to split lines in a paragraph in a way that makes interword spacing neither too large nor too small. It first tries to break lines without hyphenation. If this fails, TEX hyphenates some words and tries again. However, even this may fail: some text simply cannot be typeset properly. This happens especially if

- the text is typeset in narrow columns
- a line contains many long words that TEX does not know how to hyphenate
- hyphenation is suppressed (by a tie-in '~' or `\hbox`) in too many places on the same line

There are two ways to compromise:

- make word spacing larger than usual
- allow some words to stick out into the margin

The second approach is the default in TEX. A black rectangle is typeset at the end of a line that sticks out into the margin.

EXAMPLE 3.31 ▷ Typesetting text newspaper-style often results in `overfull` error messages. If you enter

```
\parshape 1 0in 2.5in {\sc Silicon Valley, CA, Oct.~13.}
Home Computers Inc. introduced today an IBM-compatible
clone that runs at 120Mhz and is also compatible with the
Cray XMP and the TRS-80.  The Tyrannosaurus Model T is
available immediately and retails for \$99.95.
```

you will get

> SILICON VALLEY, CA, OCT. 13.
> Home Computers Inc. introduced today
> an IBM-compatible clone that runs at 120Mhz
> and is also compatible with the Cray XMP
> and the TRS-80. The Tyrannosaurus Model
> T is available immediately and retails for
> $99.95.

and the following error messages:

```
Overfull \hbox (16.76984pt too wide) in paragraph at lines 1--4
\tenrm an IBM-compatible clone that runs at 120Mhz
          ...
Overfull \hbox (0.15866pt too wide) in paragraph at lines 1--4
\tenrm and is also com-pat-i-ble with the Cray XMP
          ...
Overfull \hbox (8.14943pt too wide) in paragraph at lines 1--4
\tenrm and the TRS-80. The Tyran-nosaurus Model
          ...
```

An `overfull` message is printed for each line that could not be typeset properly. "`lines 1--4`" refers to the offending lines of the *source* file. "`... too wide`" refers to the amount of space (in printer points) which had to be borrowed from the margin. The second line of these error messages gives the current font (`\tenrm`) and the words that cause the problem. The typeset output has a black bar at the end of each overfull line.

This example would cause no problems if typeset in a wider column or in a smaller typeface. Usually, however, this approach is unacceptable—most publications have a fixed column width and typeface.

There are two ways to correct the problem:

- fiddling with the text. This will give the best results, but it is more complicated.
- relaxing the spacing requirements (`\tolerance`). This is not as good, but it is faster, and may often be acceptable.

To use the first approach, check the first error message and look for ways to change the text in order to make more line breaks possible. In most cases the last word ("`120Mhz`," in this example) causes the problem. It can also be entered as "`120␣Mhz`", thus allowing for an extra line break. If you make this change and typeset the text again, you will get

> SILICON VALLEY, CA, OCT. 13.
Home Computers Inc. introduced today
an IBM-compatible clone that runs at 120
Mhz and is also compatible with the Cray
XMP and the TRS-80. The Tyrannosaurus
Model T is available immediately and re-
tails for $99.95.

and the following error message

```
Overfull \hbox (4.39832pt too wide) in paragraph at lines 1-4
\tenrm XMP and the TRS-80. The Tyran-nosaurus
```

which shows that the change solved part of the problem. Now the problem lies with "`Tyran-nosaurus`." Notice that this word was not completely hyphenated. Adding

```
\hyphenation{Ty-ran-no-sau-rus}
```

at the beginning of the source file results in

> SILICON VALLEY, CA, OCT. 13.
Home Computers Inc. introduced today
an IBM-compatible clone that runs at 120
Mhz and is also compatible with the Cray
XMP and the TRS-80. The Tyrannosau-
rus Model T is available immediately and
retails for $99.95.

and no error messages.

The correct hyphenation is "Ty-ran-no-saur-us," but TeX does not allow just the last two letters of a word to be separated, since such words usually look awkward. This default can be changed with the `\righthyphenmin` command, see §3.10. Newspapers often mis-hyphenate words, due to the constraints imposed by narrow columns.

Here are some guidelines:

1. Examine the hyphenated words (in the error message) and check whether all the valid hyphenation breaks were found. If not, use the `\hyphenation` command to give additional hyphenation options.
2. Check the offending paragraph for unnecessary tie-ins or `\hbox` commands. Removing them creates more line break options.
3. See if the text may be rephrased.

As mentioned above, another way to correct overfull lines is to relax spacing requirements. This is done by using the `\tolerance` command, which will be explained after a few examples.

EXAMPLE 3.32 ▷ Typing

```
\tolerance 10000 \parshape 1 0in 2.5in {\sc Silicon
Valley, CA, Oct.~13.} Home Computers Inc. introduced today
an IBM-compatible clone that runs at 120Mhz and is also
compatible with the Cray XMP and the TRS-80.  The
Tyrannosaurus Model T is available immediately and retails
for \$99.95.
```

produces

> SILICON VALLEY, CA, OCT. 13.
> Home Computers Inc. introduced today
> an IBM-compatible clone that runs at
> 120Mhz and is also compatible with the
> Cray XMP and the TRS-80. The Ty-
> rannosaurus Model T is available imme-
> diately and retails for $99.95.

As you can see, the spacing in this example is larger than normal. The spacing will be still larger in a 2-inch column (`\parshape 1 0in 2in`):

> SILICON VALLEY, CA, OCT. 13. Home Computers Inc. introduced today an IBM-compatible clone that runs at 120Mhz and is also compatible with the Cray XMP and the TRS-80. The Tyrannosaurus Model T is available immediately and retails for $99.95.

This time you will also see `underfull` messages:

```
Underfull \hbox (badness 1817) in paragraph at lines 1-4
\sc Oct. 13. \tenrm Home Com-put-ers
                ...
Underfull \hbox (badness 1137) in paragraph at lines 1-4
\tenrm compatible clone that runs at
                ...
Underfull \hbox (badness 1365) in paragraph at lines 1-4
\tenrm with the Cray XMP and the
                ...
Underfull \hbox (badness 2005) in paragraph at lines 1-4
\tenrm TRS-80. The Ty-ran-no-sau-rus
...
```

An `underfull` message is given for each line of the output with unusually bad spacing. As with `overfull` messages, `lines 1-4` refers to the problem lines in the source file and the offending text is shown in the second line of each warning message. TEX assigns a value called *badness* to each line it typesets. The badness ranges from 0 to 10000, with 0 corresponding to a perfect line and 10000 to an awful line. The badness is computed by a complicated formula, which takes into account word spacing, use of hyphenation and a few other factors. TEX considers the line that starts with "TRS-80" as the worst (badness 2005), followed by the line "OCT. 13..." (badness 1817). Only four lines of the example were flagged, but the badness was computed for all nine (the numbers are 783, 1817, 62, 1137, 308, 1365, 2005, and 3). By default, only lines with badness more than 1000 are flagged. The `\hbadness` command is used to change this default.

SYNTAX ▷ `\hbadness` [=] ⟨number⟩

where ⟨number⟩ is between 0 and 10000. Only lines whose badness exceeds ⟨number⟩ are flagged. This command affects only the part of the document that follows it.

EXAMPLE 3.33 ▷ If you specify

```
\hbadness = 10000
```

no lines will be flagged and there will be no `underfull` messages. (The badness cannot exceed 10000.) If you set

```
\hbadness = 0
```

then all lines will be flagged and their badness will be displayed. The default setting is

```
\hbadness = 1000
```

The value specified by the `\tolerance` command is used to decide whether to typeset underfull or overfull lines.

SYNTAX ▷ `\tolerance` [=] ⟨number⟩

where ⟨number⟩ lies between 0 and 10000.

Depending on how the badness compares with the values set by `\tolerance` and `\hbadness`, there are three possibilities:

1. If the badness of a line exceeds `\tolerance`, an overfull line will be typeset with a black bar on the margin and you will get an `overfull` error message.
2. If the badness is less than or equal to `\tolerance`, but greater than `\hbadness`, an underfull line will be typeset and you will get an `underfull` warning message.
3. If the badness is less than or equal to both `\tolerance` and `\hbadness`, the line will be considered acceptable.

Notice that the second case never occurs if `\tolerance` is less than `\hbadness`. The default `\tolerance` is 200 and the default `\hbadness` is 1000. This means that no underfull lines will be typeset; a line will be accepted if its badness is less than or equal to 200 and will be flagged as overfull otherwise.

If you set

```
\tolerance=10000
```

no line will be overfull (case 1). Instead, some lines will be typeset as underfull, with large spacing. Lines with badness exceeding 1000 will cause underfull messages. These messages can be suppressed with

```
\hbadness=10000
```

However, this is not advisable because you would then not know which lines are bad.

When preparing a rough draft of a document, it may be useful to set `\tolerance` equal to a large value (1000, for example). This suppresses most of the overfull lines, allowing you to concentrate on the writing, without having to deal with typesetting details. A larger value (3000–4000) for `\tolerance` is generally required for text typeset in narrow columns, because such text tends to be badly spaced (see examples above).

There is yet one more way to control the message reporting on the badly formatted lines. TEX allows lines in a paragraph to be of slightly uneven length. In fact, no overfull error is reported if the excess length of a line is under one tenth of a point. You can specify greater allowance by changing the `\hfuzz` parameters.

SYNTAX ▷ `\hfuzz` [=] ⟨dimen⟩

Finally, there are also analogs of `\hbadness` and `\hfuzz` for the page-breaking operations. These are called `\vbadness` and `\vfuzz`.

3.12 Vertical Spacing

TOPICS ▷ `\baselineskip`, `\parskip`

Most of the formatting commands discussed so far deal with composing lines from words. These commands are part of the *horizontal mode* of operation. There is also a *vertical mode*, which handles composing pages from lines. The two modes are somewhat parallel: some horizontal commands have vertical counterparts (for example, `\vskip` and `\hskip`), line breaks correspond to page breaks, underfull pages correspond to underfull lines, etc. More details on modes can be found in §8.1.

Line spacing is based on the value specified by the `\baselineskip` command. This command sets the distance between the base lines of two consecutive lines in the document. (The base line in a printed text is an imaginary line that passes through the bottoms of letters that do not have descenders, i.e. letters like *a*, *b*, *c*, but not *g*, *q*, *y*).

SYNTAX ▷ `\baselineskip` [=] ⟨dimen⟩

(See §5.3 for the list of ⟨dimen⟩ units.) `\baselineskip` is set to 12 points by default, which corresponds roughly to a single-spaced document. To typeset a double-spaced document, set `\baselineskip` to 24 points:

```
\baselineskip=24pt
```

Of course, any intermediate values are also allowed.

The `\parskip` command is used to specify extra spacing to be inserted before each paragraph.

SYNTAX ▷ `\parskip [=]` ⟨dimen⟩

See §5.3 for the list of ⟨dimen⟩ units.

EXAMPLE 3.34 ▷ If you set

```
\baselineskip 16pt \parskip 4pt
```

the spacing between lines inside a paragraph will be 16 points and between paragraphs will be 20 points.

See §5.4–§5.7 for the description of more general forms of these commands.

3.13 Vertical Skips

TOPICS ▷ `\vskip`, `\medskip`, `\bigskip`, `\smallskip`
▷ `\medbreak`, `\bigbreak`, `\smallbreak`
▷ `\medskipamount`, `\bigskipamount`, `\smallskipamount`

Vertical skips are used to override the default spacing by specifying the desired amount of space in either absolute or font-dependent units.

To specify the exact amount of vertical skip, use `\vskip`.

SYNTAX ▷ `\vskip` ⟨dimen⟩

For example, `\vskip 1in` skips exactly one inch of space. The `\vskip` command can be used to reserve space for cut-out pictures or diagrams. It is not recommended for controlling the space after or before titles or between paragraphs.

See §5.4–§5.7 for the description of more general forms of this command.

The `\smallskip`, `\medskip`, and `\bigskip` commands skip predefined amounts ("small", "medium" and "big") of vertical space. When possible, these commands should be used instead of the absolute `\vskip` command to ensure uniform spacing throughout the document.

SYNTAX ▷ `\smallskip`
▷ `\medskip`
▷ `\bigskip`

EXAMPLE 3.35 ▷ Typing

```
First sentence. \smallskip
Second sentence. \medskip
Third  sentence. \bigskip
Fourth sentence.
```

produces

First sentence.
Second sentence.

Third sentence.

Fourth sentence.

You may use `\bigskip` (or `\bigbreak`, see below) before section titles, `\medskip` after section titles, and `\smallskip` before and after tables. Using these commands consistently ensures uniform treatment of all titles and tables in a document.

Use `\smallskipamount`, `\medskipamount`, and `\bigskipamount` to set the amount of small, medium, and large skips.

SYNTAX ▷ `\smallskipamount [=]` ⟨dimen⟩
▷ `\medskipamount [=]` ⟨dimen⟩
▷ `\bigskipamount [=]` ⟨dimen⟩

See §5.3 for the list of ⟨dimen⟩ units. By default, small skip is about 3 points, medium skip is about 6 points and big skip is about 12 points. Additional details can be found in §5.7.

EXAMPLE 3.36 ▷ Setting

```
\bigskipamount = 20pt
```

will increase the spacing before each section title, and anywhere else the `\bigskip` command is used, to 20 points.

The `\medbreak`, `\smallbreak`, and `\bigbreak` commands skip exactly as much space as `\medskip`, `\smallskip`, and `\bigskip`. `Break` commands should be used instead of `skip` commands wherever page breaks are permitted.

SYNTAX ▷ `\smallbreak`
▷ `\medbreak`
▷ `\bigbreak`

EXAMPLE 3.37 ▷ A section header might be formatted as follows:

```
\bigbreak
\centerline{\bf Section IX. Zombies and related species.}
\medskip
```

The `\bigbreak` command is used instead of `\bigskip` before the header but the `\medskip` command is used after it. This ensures that a page break will not occur immediately after the title.

3.14 Page Breaks

TOPICS ▷ `\eject`, `\page`, `\nobreak`, `\goodbreak`, `\filbreak`

Often it is necessary to control the selection of page breaks to ensure that they do not occur after a title, between a table and its legend, or before the last line of a section.

There are two ways to control the page-breaking mechanism:

- by explicitly specifying page breaks
- by telling TeX about good and bad places for page breaks but leaving the final decision to the program

The `\eject` command causes an immediate page break. It is used to ensure that the material that follows starts at the beginning of a page.

SYNTAX ▷ `\eject`

➡ The page preceding an `\eject` is usually short and contains a large amount of white space. By default, this extra space will be evenly distributed between the lines of this page. To keep the text at the top of the page and the extra space at the bottom, use the

```
\page
```

command instead of `\eject`. The following examples apply equally to both `\page` and `\eject`.

EXAMPLE 3.38 ▷ `\page` may be used before chapter headers if the chapters in the document should start on new pages:

```
\page \rightline{\bf Chapter 9} \medskip
```

➡ Do not use the `\page` command to correct bad page breaks produced by TeX (especially in large documents). Imagine that a page break occurred immediately after a section header. You could place `\page` before the header to remove the bad page break; however this would probably cause problems later. For example, if you inserted some additional text anywhere before `\page`, this command would create new kinds of problems – possibly a page with a single (widow) line. You could now remove `\page` but the next revision of the document might resurrect the original problem.

The correct approach is not to select the page breaks yourself, but specify where they are allowed and leave the rest to TeX. This will ensure that page breaks always occur in reasonable places and will relieve you from checking them after each change of the document.

The `\nobreak` and `\goodbreak` commands specify correspondingly bad and good places for page breaks.

SYNTAX ▷ `\nobreak`
▷ `\goodbreak`

EXAMPLE 3.39 ▷ Typing

```
Before opening the container, verify that it contains:
\nobreak
\list 1 Monster-Card ver 23.1B
\list 1024 memory chips
\list 1 pair of pliers
\list The installation manual
 (the one you are reading now)
\goodbreak
Now you are ready to install the Monster-Card:
```

ensures that there will be no page break before the first item of the list; page breaks are encouraged after the end of the list.

The `\filbreak` command is used to force a page break, unless there is enough space on the page to fit the text until the next `\filbreak` command.

SYNTAX ▷ `\filbreak`

For example, if you issue `\filbreak` at the end of each paragraph, the page breaks will never occur in the middle of a paragraph.

The `\smallbreak`, `\medbreak`, and `\bigbreak` commands combine breaking instructions with vertical skipping. They are discussed together with other vertical skips (see §3.13).

3.15 Footnotes

TOPICS ▷ `\footnote`, `\vfootnote`, `\nfootnote`, `\footcount`, `\footins`

TeX supports two forms of footnotes, distinguishable by their tags. The `\footnote` command requires explicit specification of the tag; `\nfootnote` uses numeric tags and automatically numbers footnotes.

SYNTAX ▷ `\footnote` ⟨tag⟩ ⟨footnote text⟩

EXAMPLE 3.40 ▷ Typing

```
Strawberry trees grow only in polar regions\footnote\dag{
North of $91^\circ$ latitude}. They\footnote*{trees}
also grow underwater.
```

results in

Strawberry trees grow only in polar regions†. They* also grow underwater.

and the two footnotes that appear below.

➡ By default, footnotes are typeset in the same font as the text. It is customary, however, to typeset them in an eight-point font. This can be accomplished by defining an eight-point font (say, `\eightpoints`) and defining each footnote as

```
\footnote*{\eightpoints...
```

which will work as long as the footnote does not contain any font switches. If it does, each font command has to be manually altered to invoke an eight-point version of the font. Furthermore, if the footnote contains math, all math fonts have to be changed to the eight-point versions.

✔ In VTEX, this is automatically taken care with the `\fontscale` and `\Small` commands (see §11.9).

If the footnote tag contains more than one character, it should be placed in braces. Type

```
These trees\footnote{**}{Sold exclusively by HomeWare} ...
```

instead of

```
These trees\footnote**{Sold exclusively by HomeWare} ...
```

The `\vfootnote` command is identical to `\footnote` except that it does not place a tag in the text.

SYNTAX ▷ `\vfootnote` ⟨tag⟩ ⟨footnote text⟩

EXAMPLE 3.41 ▷ Typing

```
\vfootnote{\P}{This footnote is not
   referenced in the text.}
```

---

† North of 91° latitude

* trees

¶ This footnote is not referenced in the text.

produces the footnote below.

✓ The \nfootnote command is used to produce automatically numbered footnotes. It is not standard (see §11.9).

SYNTAX ▷ \nfootnote ⟨footnote text⟩

The \footcount command is used to start the numbering of footnotes.

SYNTAX ▷ \footcount [=] ⟨number⟩

If you want to number the footnotes separately for each chapter of the document, set \footcount to 1 at the beginning of each chapter. (Some stylesheets may do this automatically; see §6.5).

EXAMPLE 3.42 ▷ Typing

```
The authors of this book are reluctant to demonstrate
the {\tt footcount} command\nfootnote{This would change
the footnote numbering throughout Chapter~3.  Naturally,
there is a way to preserve the numbering; see Chapter~5.}.
```

results in

The authors of this book are reluctant to demonstrate the `footcount` command[1].

➡ By default, footnotes will not be split between pages. To limit the amount of vertical space used by footnotes on any single page to, for example, 2 inches, use the

```
\dimen\footins = 2in
```

command. Footnotes that do not fit into 2 inches will be carried over to the next page.

3.16 Verbatim Printing

TOPICS ▷ \obeylines, \obeyspaces

The \obeylines command suppresses the normal paragraph formatting: each line is viewed as a separate paragraph. This command can be used to typeset program listings and similar documents.

[1] This would change the footnote numbering throughout Chapter 3. Naturally, there is a way to preserve the numbering; see Chapter 5.

SYNTAX ▷ `\obeylines`

EXAMPLE 3.43 ▷ Typing

```
{ \obeylines\tt
1 SUM=0
2 FOR I=1 TO 1000
3   SUM=SUM + I
4   NEXT I
5 PRINT "THE ANSWER IS "; SUM
}
```

produces

```
1 SUM=0
2 FOR I=1 TO 1000
3 SUM=SUM + I
4 NEXT I
5 PRINT "The answer is "; SUM
```

To ensure correct spacing, use the `\obeyspaces` command.

SYNTAX ▷ `\obeyspaces`

EXAMPLE 3.44 ▷ Adding `\obeyspaces` to the previous example produces

```
1 SUM=0
2 FOR I=1 TO 1000
3   SUM=SUM + I
4   NEXT I
5 PRINT "The answer is "; SUM
```

➡ An important shortcoming of the TEX's approach to `\obeyspaces` is that it ignores (and drops) the leading spaces is each line. This, fortunately, is remedied with the alternative declaration of `\Obeyspaces`, see §11.9.

Chapter

4

Math Mode

4.1 Ordinary Math and Display Math

TOPICS ▷ $, $$, ordinary math mode, display math mode

Typesetting rules for mathematical expressions (*math*, for short) are different from those for ordinary text. For example, the word spacing does not apply to math since there are no words. However, extra space is usually provided around mathematical symbols such as '+' or '×'. The default font for normal typesetting is roman (`cmr10`), for math it is *Math Italics* (`cmmi10`). Because of the many differences between typesetting ordinary text and typesetting math, the latter is viewed as a special *mode* (math mode as opposed to text mode).

There are two math submodes: ordinary math and display math. Ordinary math produces expressions in the body of the text, usually within sentences. Display math creates expressions that occupy separate lines between paragraphs of text.

The dollar sign ($) is used to enter and exit the ordinary math mode; the double dollar sign ($$) is used to enter and exit the display math mode.

EXAMPLE 4.1 ▷ Typing

```
Assuming that $x=y+z$, we can conclude that $$z=x-y$$
```

produces

> Assuming that $x = y + z$, we can conclude that
>
> $$z = x - y$$

The differences between the ordinary and display math modes are minor and in this book the term *math* or *math mode* is used for both.

Some commands are available only in math mode. Among other things, they are used to typeset:

- subscripts and superscripts
- fractions
- summation signs, radicals, integrals
- matrices

A large collection of special symbols is available in math mode; special commands are needed to make use of them. The complete list of special symbols and the commands to access them are given in the Appendix 1; they will be freely used in the examples in this chapter.

4.2 Subscripts and Superscripts

TOPICS ▷ `^`, `_`, `\prime`

The caret '^' and underscore '_' characters are used to generate superscripts and subscripts, respectively.

SYNTAX ▷ ^⟨argument⟩
▷ _⟨argument⟩

where ⟨argument⟩ consists of the text to be placed in the superscript or the subscript. If ⟨argument⟩ contains more than a single character, it should be put in braces.

➡ In this chapter superscripts and subscripts are collectively called *scripts*.

EXAMPLE 4.2 ▷ Typing

```
$$ E = mc^2 $$
```

results in

$$E = mc^2$$

EXAMPLE 4.3 ▷ Typing

```
$10^2$, $a_i$, $a_{ij}$, $a_{q[7]+55}$,
 $\alpha_i^j$, $t^\circ$, $a^ij$
```

results in

10^2, a_i, a_{ij}, $a_{q[7]+55}$, α_i^j, t°, a^ij

Scripts may be used only in math mode. If scripts are needed within ordinary text, math mode should be entered temporarily.

EXAMPLE 4.4 ▷ Typing

```
Today's temperature is $77^\circ$F or $25^\circ$C.
```

results in

Today's temperature is 77°F or 25°C.

Scripts may have scripts of their own. It is usually necessary to use braces to avoid ambiguity. For example, the expression `$x^y^z$` is ambiguous – it could be read as either x^{yz} or x^{y^z}. This expression causes a `double superscript` warning and is typeset as x^{yz}. The correct way to enter this expression is `$x^{yz}$` if you mean x^{yz} and `$x^{y^z}$` if you mean x^{y^z}. The same applies to subscripts.

Scripts can be nested to any level. Math is normally typeset in a 10-point font, the first level scripts are typeset in a 7-point font, and the scripts on the second and subsequent levels are typeset in a 5-point font.

The roman `\rm`, slant `\sl`, italics `\it`, typewriter `\tt`, and boldface `\bf` font commands may be used in scripts.

EXAMPLE 4.5 ▷ Typing

```
$$\tau_{\rm extinction}=
  \tau_{\rm absorption}+\tau_{\rm scattering}$$
```

produces

$$\tau_{\rm extinction}=\tau_{\rm absorption}+\tau_{\rm scattering}$$

The use of other fonts in math is described in §4.3 and §4.20.

The `\prime` command produces a prime superscript. In math mode, the quote character (`'`) can be used as an abbreviation for `^\prime`.

EXAMPLE 4.6 ▷ Typing

```
$$a^\prime=b^\prime{}^\prime+c'''+{d'}^2$$
```

produces

$$a'=b''+c'''+{d'}^2$$

The empty group `{}` is used to avoid the `double superscript` warning.

The next example demonstrates some techniques for the fine alignment of scripts.

EXAMPLE 4.7 ▷ Typing

```
$F_i^j$, $F{}_i^j$, $F_i{}^j_k$, ${}^{234}U_{92}$
```

produces

F_i^j, $F{}_i^j$, $F_i{}^j_k$, ${}^{234}U_{92}$

Notice the use of the empty group `{}` to typeset a superscript *on the left.*

4.3 Mixing Text with Math

TOPICS ▷ `\␣`, `\hbox`

In addition to math fonts, such as math italics `cmmi` and math symbols `cmsy`, the following five fonts may be used directly in math mode:

- roman `\rm`
- italics `\it`
- slant `\sl`
- boldface `\bf`
- typewriter `\tt`

Other fonts need to be defined as math font families using the commands described in §4.20. Keep in mind that the spacing rules for math mode are different from those for ordinary text. The main difference is that in math, interword spacing is ignored. This applies both to math fonts (`cmmi`, `cmsy`) and to text fonts (`\rm`, `\it`, etc.) used inside math.

EXAMPLE 4.8 ▷ Typing

```
$ x^n + y^n = z^n {\rm implies that} x^n = z^n - y^n $
```

results in

$x^n + y^n = z^n \mathrm{impliesthat} x^n = z^n - y^n$

With ordinary math it is better not to insert text into math, but instead insert math into text. The spacing is correct if the previous example is entered as

```
$x^n + y^n = z^n$ implies that $x^n = z^n - y^n$
```

This approach will not work with display math, but there are two other ways to solve the problem.

The first method is to specify the spacing with the space command `\␣` or other spacing commands described in §4.14.

EXAMPLE 4.9 ▷ Typing

```
$$x^n + y^n = z^n {\rm\ implies\ that\ } x^n = z^n - y^n$$
```

results in

$$x^n + y^n = z^n \text{ implies that } x^n = z^n - y^n$$

The second approach is to use the \hbox (horizontal box) command, which invokes *restricted horizontal mode* inside math. The restricted horizontal mode is similar to the ordinary horizontal mode; the main difference is that the contents of a horizontal box may not span more than one line. Any fonts may be used in \hbox; the default font is roman.

EXAMPLE 4.10 ▷ Typing

```
$$ x^n + y^n = z^n \hbox{ \sc implies that }
   x^n = z^n - y^n $$
```

produces

$$x^n + y^n = z^n \text{ IMPLIES THAT } x^n = z^n - y^n$$

The spaces at the beginning and at the end of the \hbox argument are needed, otherwise you will get

$$x^n + y^n = z^n\text{IMPLIES THAT}x^n = z^n - y^n$$

➡ The \hbox command can also be used in the ordinary math mode. However, bear in mind that \hbox suppresses hyphenation (see §3.10).

4.4 Sums and Integrals

TOPICS ▷ \sum, \int, \limits, \nolimits

The \sum and \int commands are used to typeset summation and integration symbols ($\sum$, $\int$). The script commands '_' and '^' are used to position the limits of summation or integration.

EXAMPLE 4.11 ▷ Typing

```
\sum_{i=0}^{199} S[i] = \int_{-\infty}^{\infty} Q(x) dx
```

in ordinary math (between dollar signs $) produces

$\sum_{i=0}^{199} S[i] = \int_{-\infty}^{\infty} Q(x) dx$

In display math (between double dollar signs $$) the same expression is typeset as

$$\sum_{i=0}^{199} S[i] = \int_{-\infty}^{\infty} Q(x)dx$$

Observe the placement of the scripts in this example. You may use the `\limits` and `\nolimits` commands to change the position of the limits. The `\limits` command forces the scripts to appear above and below the operator; `\nolimits` forces the scripts to appear on the right.

EXAMPLE 4.12 ▷ Typing

```
\sum\limits_{i=0}^{\infty} F(n) =
\sum\nolimits_{i=0}^{100} G(n)
```

results in

$$\sum\limits_{i=0}^{\infty} F(n) = \sum\nolimits_{i=0}^{100} G(n)$$

in either ordinary ($) or display ($$) math.

The `\limits` and `\nolimits` commands may be applied only to math operators. In addition to `\int` and `\sum`, they include `\prod` ($\prod$), `\oint` ($\oint$) and a few more operators; the complete list is given in Appendix 1.

4.5 Vertical Stacking. Fractions

TOPICS ▷ `\atop`, `\over`, `\above`, `\choose`
▷ `\atopwithdelims`, `\overwithdelims`, `\abovewithdelims`

The commands described in this section are used to arrange vertically parts of mathematical expressions. They are used to construct fractions, fractional expressions, binomial coefficients, and simple column vectors. Some of the commands introduced in this section provide a fraction line to separate the top from the bottom; some will just stack them without a line. You may also control the thickness of the fraction line and specify delimiters (grouping symbols) such as parentheses, brackets, braces, or vertical bars '|' (see §4.8 for the list of delimiters).

All seven commands follow the same general syntax.

SYNTAX ▷ ⟨top⟩ ⟨command⟩ [options] ⟨bottom⟩

where ⟨command⟩ is one of the commands listed above. The options that apply to each command are described below. All commands view the expression on the left as ⟨top⟩ and the expression on the right as ⟨bottom⟩. Both the ⟨top⟩ and the ⟨bottom⟩ are centered.

The `\atop` command is the simplest command of this category; it places the ⟨top⟩ on top of the ⟨bottom⟩ and does not provide a fraction line.

SYNTAX ▹ ⟨top⟩ `\atop` ⟨bottom⟩

EXAMPLE 4.13 ▹ Typing `$10 \atop 19999$` produces $10 \atop 19999$

The `\over` command works in the same way as `\atop`, but it also supplies a fraction line 0.4 point thick.

SYNTAX ▹ ⟨top⟩ `\over` ⟨bottom⟩

EXAMPLE 4.14 ▹ Typing `$10 \over 19999$` produces $10 \over 19999$

The `\above` command is similar to `\over`, except that the thickness of the fraction line has to be explicitly specified.

SYNTAX ▹ ⟨top⟩ `\above` ⟨dimen⟩ ⟨bottom⟩

where ⟨dimen⟩ is the thickness of the fraction line.

EXAMPLE 4.15 ▹ Typing `$10 \above 2pt 19999$` produces $10 \above 2pt 19999$

The `\atopwithdelims` command is similar to `\atop`, but delimiters are appended to the fractional expression. The delimiters are automatically scaled.

SYNTAX ▹ ⟨top⟩ `\atopwithdelims` ⟨left-delim⟩ ⟨right-delim⟩ ⟨bottom⟩

EXAMPLE 4.16 ▹ Typing `$10 \atopwithdelims[] 19999$` produces $10 \atopwithdelims[] 19999$

The `\choose` command is a special case of `\atopwithdelims` with the delimiters being the left and right parentheses. `\choose` is primarily used for binomial coefficients.

SYNTAX ▹ ⟨top⟩ `\choose` ⟨bottom⟩

EXAMPLE 4.17 ▹ Typing `$3 \choose 1$` produces $3 \choose 1$

The `\overwithdelims` command is similar to `\over`, but delimiters are appended to the fractional expression. The delimiters are automatically scaled.

SYNTAX ▹ ⟨top⟩ `\overwithdelims` ⟨left-delim⟩ ⟨right-delim⟩ ⟨bottom⟩

EXAMPLE 4.18 ▷ `$10 \overwithdelims [] 19999$` produces $\left[{10 \over 19999}\right]$

The `\abovewithdelims` command is similar to `\above`, but delimiters are appended to the fractional expression. The delimiters are automatically scaled.

SYNTAX ▷ ⟨top⟩ `\abovewithdelims` ⟨left-delim⟩ ⟨right-delim⟩ ⟨dimen⟩ ⟨bottom⟩

EXAMPLE 4.19 ▷ `$10 \abovewithdelims [] .5pt 19999$` produces $\left[{10 \over 19999}\right]$

EXAMPLE 4.20 ▷ You may create complex fractions (fractions of fractions). Typing

```
$$ { {x+y} \over {x-y} } \over { {a+b} \over {c-d} } $$
```

results in

$$\frac{\frac{x+y}{x-y}}{\frac{a+b}{c-d}}$$

It is possible to write an ambiguous expression where it is not clear which part of the expression is supposed to be the top and which part should be viewed as the bottom. For example, if you enter

```
$$1 + 1 \over 2 + 2$$
```

you will get

$$\frac{1+1}{2+2}$$

and not

$$1+\frac{1}{2}+2$$

Usually (but not always) you will receive a warning message. Use braces to avoid ambiguity. Write `1 + { 1\over 2 } + 2` instead of `1 + 1 \over 2 + 2`.

We conclude this section with several more examples.

EXAMPLE 4.21 ▷ Typing

```
$${T_i-T_{i-1} \overwithdelims||T_i}<\varepsilon$$
```

results in

$$\left|\frac{T_i-T_{i-1}}{T_i}\right|<\varepsilon$$

EXAMPLE 4.22 ▷ Typing

```
$$|{T_i-T_{i-1} \over T_i}|<\varepsilon$$
```

produces

$$|\frac{T_i-T_{i-1}}{T_i}|<\varepsilon$$

Notice that `\overwithdelims` adjusts the size of the delimiters to match the enclosed expression. If you use `\over` and enter the delimiters separately, their size is not adjusted (see also `\left`, `\right` in §4.8).

EXAMPLE 4.23 ▷ Typing

```
$${\sin x\over x}\simeq {1-0.13356x^2+0.00328x^
4\over 1+0.03310x^2+0.00046x^4}$$
```

produces

$$\frac{\sin x}{x}\simeq\frac{1-0.13356x^2+0.00328x^4}{1+0.03310x^2+0.00046x^4}$$

In this example, the `\sin` command was used to typeset the sine function in roman instead of math italics. Similar commands are available for other commonly used functions; they are described in the next section.

4.6 Standard Function Names

TOPICS ▷ `\arccos`, `\arcsin`, `\arctan`, `\arg`, `\cos`, `\cosh`
▷ `\cot`, `\coth`, `\csc`, `\deg`, `\det`, `\dim`, `\exp`, `\gcd`
▷ `\hom`, `\inf`, `\ker`, `\lg`, `\lim`, `\liminf`, `\limsup`
▷ `\in`, `\log`, `\max`, `\min`, `\pr`, `\sec`, `\sin`
▷ `\sinh`, `\sup`, `\tan`, `\tanh`, `\bmod`, `\pmod`

By default, everything typed in math mode appears in math italics. However, it is customary to typeset the names of standard mathematical functions in roman. A special set of commands simplifies entering these names. With the exception of the `\pmod` and `\bmod` commands (see below), all commands consist of the function name preceded by a backslash.

EXAMPLE 4.24 ▷ Typing

```
$$\sin(-x) = - sin(x)$$
```

results in

$$\sin(-x) = -sin(x)$$

Using the macro facility, you can define additional commands for other functions (see §6.6).

Function names may be supplied with scripts. Depending on the function and the mode (ordinary or display), scripts are placed in the `\nolimits` position (to the right) or in the `\limits` position (above and below).

EXAMPLE 4.25 ▷ Typing

```
$$\lim_{x\rightarrow 0} {\sin(x) \over x} \rightarrow 1$$
```

results in

$$\lim_{x\rightarrow 0} \frac{\sin(x)}{x} \rightarrow 1$$

The ‘mod’ symbol is used in two ways: as a binary operation (binary mod, `\bmod`), and in a congruence statement. In the second case, ‘mod’ appears between parentheses at the end of the formula (parenthesized mod, `\pmod`).

The `\bmod` command generates a binary or infix ‘mod’.

SYNTAX ▷ `\bmod`

The `\pmod` command generates a postfix 'mod', by placing 'mod' inside parentheses at the end of an expression. The parentheses are automatically supplied.

SYNTAX ▷ `\pmod` ⟨argument⟩

EXAMPLE 4.26 ▷ Typing

```
$$1987 \bmod 25 = 12       $$
$$1987 \equiv 12 \pmod{25}$$
```

produces

$$1987 \bmod 25 = 12$$
$$1987 \equiv 12 \pmod{25}$$

4.7 Roots

TOPICS ▷ `\sqrt`, `\root`

The `\sqrt` command generates a square root symbol.

SYNTAX ▷ `\sqrt` ⟨radicand⟩

where ⟨radicand⟩ is the expression to be placed inside the radical symbol.

EXAMPLE 4.27 ▷ Typing

```
$$\sqrt{a+2b}+\sqrt{x+y \over x-y}$$
```

produces

$$\sqrt{a+2b}+\sqrt{x+y \over x-y}$$

The radical symbols are scaled to encompass the radicand.

The `\root` command generates root symbols of any index.

SYNTAX ▷ `\root` ⟨index⟩ `\of` ⟨radicand⟩

where ⟨index⟩ denotes the index of the root.

EXAMPLE 4.28 ▷ Typing

```
$$\root7\of{105}\times\root13\of{\Gamma(x+y)\over{x+y}}$$
```

produces

$$\sqrt[7]{105}\times\sqrt[13]{\frac{\Gamma(x+y)}{x+y}}$$

EXAMPLE 4.29 ▷ Radical symbols can be nested. Typing

```
$$\sqrt{1+\sqrt{2+\sqrt{3+\sqrt {4+\sqrt{5+\sqrt 6}}}}}$$
```

results in

$$\sqrt{1+\sqrt{2+\sqrt{3+\sqrt{4+\sqrt{5+\sqrt{6}}}}}}$$

➡ The radicand must be enclosed in braces unless it is a single character. For example, typing `$\sqrt a+b$` produces $\sqrt{a}+b$ and not $\sqrt{a+b}$.

4.8 Big and Adjustable Delimiters

TOPICS ▷ `\lbrack`, `\rbrack`, `\lbrace`, `\rbrace`, `\langle`, `\rangle`, `\lfloor`,
▷ `\rfloor`, `\lceil`, `\rceil`, `\vert`, `\Vert`, `\backslash`, `\uparrow`,
▷ `\downarrow`, `\Uparrow`, `\Downarrow`, `\updownarrow`, `\Updownarrow`,
▷ `\arrowvert`, `\Arrowvert`, `\bracevert`, `\lgroup`, `\rgroup`
▷ `\lmoustache`, `\rmoustache`
▷ `\bigl`, `\bigr`, `\Bigl`, `\Bigr`, `\biggl`, `\biggr`, `\Biggl`, `\Biggr`, `\bigm`,
▷ `\Bigm`, `\biggm`, `\Biggm`, `\big`, `\Big`, `\bigg`, `\Bigg`, `\left`, `\right`

The basic delimiters are listed in the following table:

<table>
<tr><th colspan="3">Basic Delimiters</th></tr>
<tr><th>Symbol</th><th>Command</th><th>Name</th></tr>
<tr><td>(</td><td><code>(</code></td><td>left parenthesis</td></tr>
<tr><td>)</td><td><code>)</code></td><td>right parenthesis</td></tr>
<tr><td>[</td><td><code>[</code> or <code>\lbrack</code></td><td>left bracket</td></tr>
<tr><td>]</td><td><code>]</code> or <code>\rbrack</code></td><td>right bracket</td></tr>
<tr><td>{</td><td><code>\{</code> or <code>\lbrace</code></td><td>left curly brace</td></tr>
<tr><td>}</td><td><code>\}</code> or <code>\rbrace</code></td><td>right curly brace</td></tr>
<tr><td>⟨</td><td><code>\langle</code></td><td>left angle bracket</td></tr>
<tr><td>⟩</td><td><code>\rangle</code></td><td>right angle bracket</td></tr>
<tr><td>⌊</td><td><code>\lfloor</code></td><td>left floor bracket</td></tr>
<tr><td>⌋</td><td><code>\rfloor</code></td><td>right floor bracket</td></tr>
<tr><td>⌈</td><td><code>\lceil</code></td><td>left ceiling bracket</td></tr>
<tr><td>⌉</td><td><code>\rceil</code></td><td>right ceiling bracket</td></tr>
<tr><td>|</td><td><code>|</code> or <code>\vert</code></td><td>vertical bar</td></tr>
<tr><td>‖</td><td><code>\|</code> or <code>\Vert</code></td><td>double vertical bar</td></tr>
<tr><td>/</td><td><code>/</code></td><td>slash</td></tr>
<tr><td>\</td><td><code>\backslash</code></td><td>back slash</td></tr>
<tr><td>↑</td><td><code>\uparrow</code></td><td>upward arrow</td></tr>
<tr><td>↓</td><td><code>\downarrow</code></td><td>downward arrow</td></tr>
<tr><td>⇑</td><td><code>\Uparrow</code></td><td>double upward arrow</td></tr>
<tr><td>⇓</td><td><code>\Downarrow</code></td><td>double downward arrow</td></tr>
<tr><td>↕</td><td><code>\updownarrow</code></td><td>up-and-down arrow</td></tr>
<tr><td>⇕</td><td><code>\Updownarrow</code></td><td>double up-and-down arrow</td></tr>
</table>

Delimiters often have to be adjusted to the height of the expression they enclose. In many cases their size is automatically computed, as with `\overwithdelims`. There are two ways in which you can control the scaling of delimiters.

The first approach is to choose one of four basic sizes. In increasing order they are "big", "Big", "bigg", and "Bigg".

- `\bigl`, `\Bigl`, `\biggl`, and `\Biggl` scale left delimiters.
- `\bigr`, `\Bigr`, `\biggr`, and `\Biggr` scale right delimiters.
- `\bigm`, `\Bigm`, `\biggm`, and `\Biggm` scale interior delimiters such as a slash that denotes division.

EXAMPLE 4.30 ▷ Typing

```
$$*\Biggl(*\biggl(*\Bigl(*\bigl(*\bigr)
  *\Bigr)*\biggr)*\Biggr)*$$
```

results in

$$*\Biggl(*\biggl(*\Bigl(*\bigl(*\bigr)*\Bigr)*\biggr)*\Biggr)*$$

➡ You must specify whether the delimiter is right (`r`), left (`l`) or middle (`m`) in order to ensure correct spacing around it. It is not an error to enter `\bigl )`. This will produce a `\big` sized right parenthesis, but the spacing will be that of a left parenthesis. If spacing is not important you may use the `\big`, `\Big`, `\bigg`, or `\Bigg` commands.

EXAMPLE 4.31 ▷ Typing

```
$$ax^3+bx^2+cx+d =
  d + \Bigl(c+ \bigl(b+ (a)x \bigr)x \Bigr) x$$
$$\bigl (y \in f(x) \bigm\vert y \in g(x) \bigr) $$
```

results in

$$ax^3+bx^2+cx+d = d + \Bigl(c+ \bigl(b+ (a)x \bigr)x \Bigr) x$$

$$\bigl(y \in f(x) \bigm\vert y \in g(x) \bigr)$$

The second approach is to use the `\left` and `\right` commands.

SYNTAX ▷ `\left` ⟨left-delim⟩ ⟨string⟩ `\right` ⟨right-delim⟩

`\right`'s should match `\left`'s. The size of the delimiters corresponds to the height of the enclosed expression.

EXAMPLE 4.32 ▷ Typing

```
$$ 2\times \left[ {\left(x+y\over x+1\right)^2}
   + \sqrt{z+0.23} \right] $$
```

produces

$$2 \times \left[{\left(x+y\over x+1\right)^2} + \sqrt{z+0.23} \right]$$

The delimiters used with matched `\left` and `\right` commands need not be of the same kind, nor do they have to be of the proper kind; for example, you may use a left bracket with `\right`.

EXAMPLE 4.33 ▷ Typing

```
According to Dieudonn\'e,
$ x \in \left[ {n - 1 \over n}, {n + 1 \over n} \right) $
should be written as
$ x \in \left[ {n - 1 \over n}, {n + 1 \over n} \right[ $.
```

produces

According to Dieudonné, $x \in \left[{n-1 \over n}, {n+1 \over n} \right)$ should be written as $x \in \left[{n-1 \over n}, {n+1 \over n} \right[$.

There is also a dummy (null) delimiter, which is entered as a period. Use it if you need only one delimiter.

EXAMPLE 4.34 ▷ Typing

```
$$f(x)=\left\{x^2, {\rm\ if\ } x>0 \atop x^3,
 {\rm\ otherwise\ } \right.$$
```

produces

$$f(x)=\left\{ {x^2, \text{ if } x>0 \atop x^3, \text{ otherwise }} \right.$$

The `\cases` command discussed in §4.12 provides a different (and a simpler) way to typeset this example. Using `\cases` it is also possible to align the right hand side.

There are seven more delimiters that are available only in large sizes ("Big", "bigg", or "Bigg"):

- `\arrowvert` : $\Big\arrowvert$
- `\Arrowvert` : $\Big\Arrowvert$
- `\bracevert` : $\Big\bracevert$
- `\lgroup` and `\rgroup` : $\Big\lgroup\Big\rgroup$
- `\lmoustache` and `\rmoustache` : $\Big\lmoustache\Big\rmoustache$

All the examples above are shown in the "Big" size.

4.9 Math Styles

TOPICS ▷ `\displaystyle`, `\textstyle`, `\scriptstyle`, `\scriptscriptstyle`

Two major differences between ordinary math ($) and display math ($$) are spacing and choice of character size. The same expression often looks different in each mode. For example,

```
{\sqrt{17}+2^2 \over 53} + \sum_{i=0}^{1987} \psi(x^2)
```

appears in display math as

$$\frac{\sqrt{17}+2^2}{53} + \sum_{i=0}^{1987} \psi(x^2)$$

and in ordinary math as

$\textstyle \frac{\sqrt{17}+2^2}{53} + \sum_{i=0}^{1987} \psi(x^2)$

If the entire expression appears in a first-level script, it becomes

$\scriptstyle \frac{\sqrt{17}+2^2}{53} + \sum_{i=0}^{1987} \psi(x^2)$

In a higher-level script it has still another form:

$\scriptscriptstyle \frac{\sqrt{17}+2^2}{53} + \sum_{i=0}^{1987} \psi(x^2)$

Expressions are typeset in four styles:

- display style for display math
- text style for ordinary math
- script style for first-level scripts
- scriptscript style for higher-level scripts

The `\displaystyle` command forces an expression to be typeset in display style, regardless of whether the expression appears in display math or in ordinary math or in a script. Similarly, `\textstyle` forces an expression to be typeset as in ordinary math, `\scriptstyle` as in a first-level script, and `\scriptscriptstyle` as in a higher-level script.

The examples above were typeset by using the appropriate style command. The first was entered as

```
${ \displaystyle {\sqrt{17}+2^2 \over 53}
+ \sum_{i=0}^{1987} \psi(x^2)} $
```

In the other three the `\textstyle`, `\scriptstyle`, and `\scriptscriptstyle` commands were used instead of `\displaystyle`.

4.10 Accents, Bars, Hats

TOPICS ▷ `\acute`, `\bar`, `\breve`, `\check`, `\ddot`, `\dot`, `\grave`, `\hat`
▷ `\tilde`, `\vec`, `\skew`, `\widetilde`, `\widehat`

The `\acute`, `\bar`, `\breve`, `\check`, `\ddot`, `\dot`, `\grave`, `\hat`, `\tilde`, and `\vec` commands affix accents to symbols in math mode. These commands have different names from their text mode counterparts (see §2.12), since they are intended for different purposes. The table below illustrates their use

Math Accents		
In Math	In Text	Example
`\acute`	`\'`	$\acute{X}$
`\bar`	`\=`	$\bar{X}$
`\breve`	`\u`	$\breve{X}$
`\check`	`\v`	$\check{X}$
`\ddot`	`\"`	$\ddot{X}$
`\dot`	`\.`	$\dot{X}$
`\grave`	``\` ``	$\grave{X}$
`\hat`	`\^`	$\hat{X}$
`\tilde`	`\~`	$\tilde{X}$
`\vec`	none	$\vec{X}$

Math accents, unlike text accents, can be nested. For example, `$\acute{\hat X}$` is typeset as $\acute{\hat X}$. The `\skew` command shifts the position of the accent.

SYNTAX ▷ `\skew` ⟨number⟩

EXAMPLE 4.35 ▷ Typing

```
$$ {\skew1\bar{\bar X}}
   {\skew2\bar{\bar X}}
        ...
   {\skew{10}\bar{\bar X}} $$
```

results in

$${\skew1\bar{\bar X}}{\skew2\bar{\bar X}}{\skew3\bar{\bar X}}{\skew4\bar{\bar X}}{\skew5\bar{\bar X}}{\skew6\bar{\bar X}}{\skew7\bar{\bar X}}{\skew8\bar{\bar X}}{\skew9\bar{\bar X}}{\skew{10}\bar{\bar X}}$$

The optimal position is best found through trial and error.

Math accents should be applied only to a single character. If you apply an accent to a group of characters, it will be placed over the center of the group; the size of the accent will not be adjusted. The `\widetilde` and `\widehat` commands produce partially adjustable accents.

EXAMPLE 4.36 ▷ Typing

```
$$ \ \tilde{x}     \ \tilde{xy}     \ \tilde{xyz}
   \ \widetilde{x} \ \widetilde{xy} \ \widetilde{xyz}
   \ \hat{x}       \ \hat{xy}       \ \hat{xyz}
   \ \widehat{x}   \ \widehat{xy}   \ \widehat{xyz} $$
```

results in

$$\tilde{x}\ \tilde{xy}\ \tilde{xyz}\ \widetilde{x}\ \widetilde{xy}\ \widetilde{xyz}\ \hat{x}\ \hat{xy}\ \hat{xyz}\ \widehat{x}\ \widehat{xy}\ \widehat{xyz}$$

Both `\widehat` and `\widetilde` come in three sizes: for one, two, or three characters.

The wide versions of the `\bar` and `\vec` accents are `\overline` and `\overrightarrow` (see §4.11).

4.11 Lines, Arrows, Captions

TOPICS ▷ `\mathstrut`, `\overline`, `\overbrace`, `\underline`, `\underbrace`
▷ `\overrightarrow`, `\overleftarrow`, `\buildrel`, `\joinrel`

The `\overline` and `\underline` commands do what their names suggest: draw a rule over or under a group of characters.

SYNTAX ▷ `\underline` ⟨argument⟩
▷ `\overline` ⟨argument⟩

EXAMPLE 4.37 ▷ Typing

```
$\overline {a + bi} = a - bi$
```

produces

$\overline{a+bi} = a - bi$

EXAMPLE 4.38 ▷ Typing

```
$$ 3ab + \underline{2ac} + \underline{\underline{bc}} -
\underline{5ac}-
   \underline{\underline{7bc}}=3ab-6bc-3ac$$
```

produces

$$3ab + \underline{2ac} + \underline{\underline{bc}} - \underline{5ac} - \underline{\underline{7bc}} = 3ab - 6bc - 3ac$$

The position of the rule depends on the height or the depth of the characters that are overlined or underlined.

EXAMPLE 4.39 ▷ Typing

```
$$\overline  b  + \overline e + \overline  g +
  \underline x + \underline y + \underline \phi $$
```

produces

$$\overline b + \overline e + \overline g + \underline x + \underline y + \underline \phi$$

The `\mathstrut` command is used to align all rules within an expression.

SYNTAX ▷ `\mathstrut` ⟨argument⟩

EXAMPLE 4.40 ▷ Typing

```
$$\overline {\mathstrut b}
 +\overline {\mathstrut e}+\overline {\mathstrut g}
 +\underline{\mathstrut x}+\underline{\mathstrut y}
 +\underline{\mathstrut \phi}$$
```

results in

$$\overline {\mathstrut b} +\overline {\mathstrut e}+\overline {\mathstrut g} +\underline{\mathstrut x}+\underline{\mathstrut y} +\underline{\mathstrut \phi}$$

The `\overbrace` and `\underbrace` commands generate braces above and below an expression. The `\overrightarrow` and `\overleftarrow` commands generate arrows pointing respectively right and left above an expression.

EXAMPLE 4.41 ▷ Typing

```
$$\overbrace{abc}    \ \ \underbrace{0123456789}, \ \
\overrightarrow{abc}\ \ \overleftarrow{0123456789}$$
```

results in

$$\overbrace{abc} \ \ \underbrace{0123456789}, \ \ \overrightarrow{abc}\ \ \overleftarrow{0123456789}$$

You can use script commands to add captions to braces and arrows.

EXAMPLE 4.42 ▷ Typing

```
$$letters=\{ \underbrace{a,b,c,\ldots,z,A,B,\ldots,Z}_%
   {52\rm\ characters}\}$$
```

results in

$$letters = \{\underbrace{a, b, c, \ldots, z, A, B, \ldots, Z}_{\rm 52\ characters}\}$$

(The `\ldots` command is described in §4.13.)

The `\buildrel` command can be used to place symbols over horizontal arrows.

SYNTAX ▷ `\buildrel` ⟨superscript expression⟩ `\over` ⟨arrow⟩

EXAMPLE 4.43 ▷ Typing

```
$$N_2+O_2\buildrel k_p\over\longrightarrow 2NO$$
```

produces

$$N_2 + O_2 \buildrel k_p \over \longrightarrow 2NO.$$

You can build longer arrows by putting together symbols listed in Appendix 1. In the math mode the shapes of the equal sign '=' and minus '−' match the shapes of arrow symbols (e.g., `\Rightarrow`). Use the `\joinrel` command to eliminate the space between components.

SYNTAX ▷ ⟨symbol$_1$⟩ `\joinrel` ⟨symbol$_2$⟩

EXAMPLE 4.44 ▷ Typing

```
$ = = = = = = = = \Rightarrow$ \hbox{ becomes }
$ =\joinrel=\joinrel=\joinrel=\joinrel
  =\joinrel=\joinrel=\joinrel=\joinrel\Rightarrow $
```

produces

$= = = = = = = = \Rightarrow$ becomes $=\joinrel=\joinrel=\joinrel=\joinrel=\joinrel=\joinrel=\joinrel=\joinrel\Rightarrow$

Use the macro facility (see Chapter 6) to define terse commands for entering composite symbols.

4.12 Matrices

TOPICS ▷ `\matrix, \pmatrix, &, \cr, \cases`

To typeset a matrix you list its entries row by row.

SYNTAX ▷ `\matrix {` ⟨list of entries⟩ `}`

where ⟨list of entries⟩ contains the entries of the matrix, with the rows separated by the `\cr` (carriage return) command and the columns separated by the `&` (alignment tab) command. Each line must have the same number of alignment tabs.

EXAMPLE 4.45 ▷ Typing

```
$$ \matrix{1 & 2 \cr 3 & 4} $$
```

results in

$$\matrix{1 & 2 \cr 3 & 4}$$

There are no syntactical restrictions on the size of the matrix or of its entries. However, trying to typeset a very large matrix may cause the program to run out of memory.

EXAMPLE 4.46 ▷ Typing

```
$$\matrix{1&2\over755&19.22312
  \cr\sqrt{17}\over3&\sin(x)&1}$$
```

results in

$$\matrix{1 & \frac{2}{755} & 19.22312 \cr \frac{\sqrt{17}}{3} & \sin(x) & 1}$$

The `\pmatrix` command is similar to `\matrix`; in addition, it encloses the matrix between parentheses of the right size.

SYNTAX ▷ `\pmatrix {` ⟨list of entries⟩ `}`

EXAMPLE 4.47 ▷ Typing

```
$$\pmatrix{ 1 & a & a^2  \cr
            1 & b & b^2  \cr
            1 & c & c^2      } $$
```

results in

$$\begin{pmatrix} 1 & a & a^2 \\ 1 & b & b^2 \\ 1 & c & c^2 \end{pmatrix}$$

➡ Although TEX ignores the formatting in the source file, it is good practice to type matrix specifications in a structured way.

There are no restrictions on the complexity of each entry—it may contain text or other matrices.

EXAMPLE 4.48 ▷ Typing

```
\pmatrix{ \matrix {1 & 2 \cr 3 & 4}     &
          \sqrt{99+\sqrt{8} \over 33 } \cr
          \hbox{(A true monster)}       &
          \pmatrix{1&2\cr 3&4\cr 5&6}  }
```

produces

$$\begin{pmatrix} \begin{matrix} 1 & 2 \\ 3 & 4 \end{matrix} & \sqrt{\frac{99+\sqrt{8}}{33}} \\ \text{(A true monster)} & \begin{pmatrix} 1 & 2 \\ 3 & 4 \\ 5 & 6 \end{pmatrix} \end{pmatrix}$$

Entries may be left blank by simply omitting them but including the alignment tabs.

EXAMPLE 4.49 ▷ Typing

```
$$\pmatrix{ 1 & 2 & 3 \cr 4 & & \cr 7 & 8 & 9}$$
```

produces

$$\begin{pmatrix} 1 & 2 & 3 \\ 4 & & \\ 7 & 8 & 9 \end{pmatrix}$$

The `\pmatrix` command works like `\matrix` placed between "`\left(`" and "`\right)`". Other stretchable delimiters (see §4.8) may also be used.

EXAMPLE 4.50 ▷ For instance, typing

```
$$\left\vert
\matrix{ \cos(\phi) & \sin(\phi) \cr
   \sin(-\phi) & \cos(\phi) }
```

```
\right\vert $$
```

produces

$$\left|\begin{matrix}\cos(\phi) & \sin(\phi)\\ \sin(-\phi) & \cos(\phi)\end{matrix}\right|$$

➡ If you have to enter determinants repeatedly, it is best to define a macro for this purpose (see §6.6).

The `\cases` command is used to typeset a list of possibilities.

SYNTAX ▷ `\cases {` ⟨case list⟩ `}`

⟨case list⟩ is entered as a two-column matrix. The second column is typeset in text mode, since it is normally used to provide descriptions for each case. Each entry in the second column is typeset as if it were placed in an "`\hbox`" (see §4.3).

EXAMPLE 4.51 ▷ Typing

```
The Banana Islands Sultanate computes
  income tax as follows:
$$ Tax = \cases {0\% & for members of the royal family \cr
                 5\% & for their friends                \cr
                 90\% & for everyone else  }        $$
```

results in

The Banana Islands Sultanate computes income tax as follows:

$$Tax = \begin{cases} 0\% & \text{for members of the royal family} \\ 5\% & \text{for their friends} \\ 90\% & \text{for everyone else} \end{cases}$$

Ordinary math mode ($) must be re-entered for math symbols that appear in the second column.

EXAMPLE 4.52 ▷ Typing

```
$$ abs(x) = \cases{ x  & $x$ is positive \cr
                    0  & $x=0$           \cr
                   -x  & $x$ is negative } $$
```

results in

$$abs(x) = \begin{cases} x & x \text{ is positive} \\ 0 & x = 0 \\ -x & x \text{ is negative} \end{cases}$$

By default, the left column of `\cases` is left-justified. However, the example above would look better if it were right-justified. VTEX predefines simple macros `\rj` and `\lj` for use with `\cases` and `\matrix` (see §11.9 for details).

EXAMPLE 4.53 ▷ Typing

```
$$ abs(x) = \cases{ \rj{x}  & $x$ is positive \cr
                    \rj{0}  & $x=0$           \cr
                   \rj{-x}  & $x$ is negative  } $$
```

produces

$$abs(x) = \begin{cases} \hfill x & x \text{ is positive} \\ \hfill 0 & x = 0 \\ -x & x \text{ is negative} \end{cases}$$

4.13 Ellipses

TOPICS ▷ `\ldots`, `\cdots`, `\vdots`, `\ddots`

The commands described in this section generate groups of three dots. They are commonly used in mathematical expressions to denote missing terms. You can arrange these dots horizontally (`\ldots`, `\cdots`), vertically (`\vdots`), or diagonally (`\ddots`).

SYNTAX ▷ `\ldots`
▷ `\cdots`
▷ `\vdots`
▷ `\ddots`

Here `l` stands for low, `c` for centered, `v` for vertical, and `d` for diagonal.

EXAMPLE 4.54 ▷ Typing

```
{\tt for} $x=1,2,3, \ldots, 999$
{\tt do}  $sum \leftarrow 1+2+3+\cdots+x$
```

produces

`for` $x = 1, 2, 3, \ldots, 999$ `do` $sum \leftarrow 1 + 2 + 3 + \cdots + x$

Diagonal and vertical dots appear mainly in matrices and tables.

EXAMPLE 4.55 ▷ Typing

```
$$\pmatrix{1      & \cdots & 100    \cr
           \vdots & \ddots & \vdots \cr
           100    & \cdots & 200     }
  \otimes
  \pmatrix{1      &    0     & \cdots & 0     \cr
           0      &    1     & \cdots & 0     \cr
           \vdots & \vdots   & \ddots &\vdots\cr
           0      &    0     & \cdots & 1     }$$
```

results in

$$\begin{pmatrix} 1 & \cdots & 100 \\ \vdots & \ddots & \vdots \\ 100 & \cdots & 200 \end{pmatrix} \otimes \begin{pmatrix} 1 & 0 & \cdots & 0 \\ 0 & 1 & \cdots & 0 \\ \vdots & \vdots & \ddots & \vdots \\ 0 & 0 & \cdots & 1 \end{pmatrix}$$

4.14 Spacing commands

TOPICS ▷ \qquad, \quad, \␣, \,, \>, \;, \!, mu

The commands described in this section allow you to alter the default spacing done by TEX.

The \quad, \qquad, and \␣ commands work in the same way as in the text mode (see §3.5). Five additional commands are provided in math mode for fine-tuning:

- \, skips a *thin* space (1/6 of a \quad)
- \> skips a *medium* space (2/9 of a \quad)
- \; skips a *thick* space (5/18 of a \quad)
- \! skips a *negative* space (−1/6 of a \quad, i.e., kerns the text 1/6 inch left)

For more precise spacing you can specify the space in special mu units. There are eighteen mu's to a quad. Thin space is, therefore, 3 mu's wide; medium space is 4 mu's, and thick space is 5 mu's. mu units follow the same syntax as other dimensional units, although they can only be used in the math mode.

EXAMPLE 4.56 ▷ Typing

```
$$c=a,\qquad d=b$$
$$c=a,\quad d=b $$
$$c=a,\; d=b $$
$$c=a,\> d=b $$
$$c=a,\, d=b $$
$$c=a,\!  d=b $$
```

produces

$$c=a,\qquad d=b$$
$$c=a,\quad d=b$$
$$c=a,\; d=b$$
$$c=a,\> d=b$$
$$c=a,\, d=b$$
$$c=a,\! d=b$$

EXAMPLE 4.57 ▷ Typing

```
$$ {\rm If\ }a=a_0, \ b=b_0 \quad {\rm then\ }
  \int\!\!\int\!\!{\int_a}^b \Gamma(\sqrt{x+y}) \ldots $$
$$ {\rm If\ }a=a_0, b=b_0 {\rm then\ }
  \int\int{\int_a}^b \Gamma(\sqrt{x+y}) \ldots      $$
```

produces

$$ {\rm If\ }a=a_0, \ b=b_0 \quad {\rm then\ } \int\!\!\int\!\!{\int_a}^b \Gamma(\sqrt{x+y}) \ldots $$
$$ {\rm If\ }a=a_0, b=b_0 {\rm then\ } \int\int{\int_a}^b \Gamma(\sqrt{x+y}) \ldots $$

4.15 Equation Numbering

TOPICS ▷ \eqno, \leqno

Mathematical equations are frequently accompanied by reference labels. The labels may be numbers or other symbols. They may be placed at right margin (\eqno) or left margin (\leqno).

SYNTAX ▷ \eqno(⟨equation label⟩)
▷ \leqno(⟨equation label⟩)

EXAMPLE 4.58 ▷ Typing

```
$$ c=a+b \eqno(8.1) $$
```

produces

$$ c=a+b \qquad (8.1) $$

EXAMPLE 4.59 ▷ Similarly, typing

```
$$ c=a+b+\sum_{i=0}^{\phi}cos({\phi\over i})\leqno(***)$$
```

produces

$$c = a + b + \sum_{i=0}^{\phi} cos(\frac{\phi}{i}) \leqno(***)$$

➡ The `\eqno` and `\leqno` commands may be used only in display math (`$$`) mode.

➡ Using the macro facility of TEX it is possible to implement automatic numbering of equations (see §6.6).

4.16 Aligning Multi-line Equations

TOPICS ▷ `\eqalign`

By default, equations typeset in display math mode (`$$`) are centered. However, in multi-line equations it is customary to align parts of equations. The `\eqalign` command is used to this end.

SYNTAX ▷ `\eqalign` ⟨argument⟩

where ⟨argument⟩ contains the equations to be aligned. The `\cr` command is used to separate expressions that should appear on different lines; there should be the same number of alignment tabs on each line.

EXAMPLE 4.60 ▷ Typing

```
$$\eqalign{\sin(2x)&= \sin(x+x)                         \cr
                   &= \sin(x)\cos(x)+\sin(x)\cos(x) \cr
                   &= 2\sin(x)\cos(x)} \eqno(17)$$
```

results in

$$\begin{aligned} \sin(2x) &= \sin(x+x) \\ &= \sin(x)\cos(x) + \sin(x)\cos(x) \\ &= 2\sin(x)\cos(x) \end{aligned} \qquad (17)$$

Compare this with the same example typeset without `\eqalign`. Typing

```
$$ \sin(2x)     = \sin(x+x)
                = \sin(x)\cos(x)+\sin(x)\cos(x)
                = 2\sin(x)\cos(x) \eqno(17) $$
```

produces

$$\sin(2x) = \sin(x + x) = \sin(x)\cos(x) + \sin(x)\cos(x) = 2\sin(x)\cos(x) \qquad (17)$$

4.17 Line Breaks in Math Mode

TOPICS ▷ `\allowbreak`, `\*`

Math expressions are generally not broken across lines. It is best to typeset a long expression in display math and not within a paragraph. When it is necessary to break a math expression, TEX proceeds conservatively: it only breaks outside all groups, i.e., not between braces and not between `\left` and `\right`, and only after a relation (e.g., '=') or an operator (e.g., '×'). If an acceptable break is not found, the paragraph becomes underfull or overfull (see §3.11).

Use the `\allowbreak` command to mark permissible line breaks.

The `\*` (*discretionary times*) command allows line breaks at the multiplication sign. For example, `$(x+1)\*(x-1)$` will be typeset as $(x+1)(x-1)$ if it fits on a single line. Otherwise, one line will end with $(x+1)\times$ and the following line will start with $(x-1)$.

4.18 Symbols. Spacing. Coercion

TOPICS ▷ `\mathord`, `\mathop`, `\mathbin`, `\mathrel`, `\mathopen`, `\mathclose`
▷ `\mathpunct`, `\mathinner`, `\colon`, `\backslash`, `\setminus`

All symbols and expressions in math mode are subdivided into eight categories:

ORD: Ordinary symbols
OP: Large operators
BIN: Binary operators
REL: Relational operators
OPEN: Opening punctuation symbols
CLOSE: Closing punctuation symbols
PUNCT: Other punctuation symbols
INNER: Fractional expressions

§11.1 contains the table of all mathematical symbols, grouped by category.

The spacing between two symbols in math mode is determined by their categories. The exact spacing rules are given by the following table:

Default Spacing Rules								
	ORD	OP	BIN	REL	OPEN	CLOSE	PUNCT	INNER
ORD	0	1	2?	3?	0	0	0	1?
OP	1	1		3?	0	0	0	1?
BIN	2?	2?			2?			2?
REL	3?	3?		0	3?	0	0	3?
OPEN	0	0		0	0	0	0	0
CLOSE	0	1	2?	3?	0	0	0	1?
PUNCT	1?	1?		1?	1?	1?	1?	1?
INNER	1?	1	2?	3?	1?	0	1?	1?

In the table, 0 corresponds to no space, 1 corresponds to thin space(`\,`), 2 corresponds to medium space (`\>`), and 3 corresponds to thick space (`\;`) in display and text styles. This spacing is also used in script styles unless the entry contains '?', in which case no space is produced. Finally, the entries left blank correspond to cases that never occur. Binary operators in equations should not follow one another. If they do, TeX forces one of them into a different category. For instance, the second plus in `a++b` will be viewed as an ordinary symbol.

You may coerce symbols of one category into another in order to modify the spacing. The coercion commands are: `\mathord`, `\mathop`, `\mathbin`, `\mathrel`, `\mathopen`, `\mathclose`, `\mathpunct`, and `\mathinner`.

These commands are needed if mathematical symbols are used in a way that differs from their normal meaning. For example, TeX views `\lfloor` '$\lfloor$' as an opening symbol. In APL (A Programming Language), '$\lfloor$' is a binary operator that computes the minimum of two numbers. If you enter `(X+9\lfloor13)` and `(X+9\mathbin{\lfloor}13)` you will get $\boxed{(X+9\lfloor 13)}$ and $\boxed{(X+9\mathbin{\lfloor}13)}$.

Similarly, the colon ':' is treated as a relational (REL) operator. To use it as a punctuation (PUNCT) symbol, use the `\colon` command. `\backslash` generates a symbol in the ORD category, while `\setminus` produces the same symbol in the BIN category. Correct commands should be used to ensure proper spacing.

4.19 Phantoms

TOPICS ▷ `\phantom`, handwritten symbols

It is sometimes necessary to make room for a symbol or group of symbols without actually typesetting them. This is accomplished with the `\phantom` command.

SYNTAX ▷ `\phantom` ⟨argument⟩

⟨argument⟩ will not appear in the output. Instead, the output will contain an empty space of the same size and shape as ⟨argument⟩, were it to be typeset there.

The `\phantom` command is often used to align entries in a matrix.

EXAMPLE 4.61 ▷ Typing

```
$$ \hbox{\rm Compare}
     \pmatrix{ 0&            1 \cr 0 & -1 }
   \hbox{\rm with}
     \pmatrix{ 0& \phantom{-}1 \cr 0 & -1 }. $$
```

produces

$$\hbox{Compare}\pmatrix{0 & 1 \cr 0 & -1}\hbox{ with }\pmatrix{0 & 1 \cr 0 & -1}.$$

✓ An alternative is to change the alignment of the entries using `\rj` and `\lj` commands (see §11.9).

The `\phantom` command may also be used to reserve space for a symbol that will be written in by hand. For best results, find an available symbol of the same category and of approximately the same size to use as ⟨argument⟩. Appendix 1 contains the table of available symbols.

EXAMPLE 4.62 ▷ Suppose that your new symbol is a math operator that is similar in size to $\prod$ (`\prod`). Use `\phantom{\prod}` wherever you are going to place your symbol. For instance, typing

```
$$ B=\mathop{\phantom{\prod}}
\limits_{i=1}^{\infty} A(i)$$
```

produces

$$B=\mathop{}\limits_{i=1}^{\infty} A(i)$$

4.20 Math Font Families

TOPICS ▷ `\textfont`, `\scriptfont`, `\scriptscriptfont`, `\mathchardef`

TeX uses a different scheme of classifying fonts in math mode. While all fonts are independent of each other in text mode, in math mode they are grouped in 16 families. Each font family consists of three fonts: `\textfont`, `\scriptfont` and `\scriptscriptfont` for typesetting text, scripts and scripts of scripts, respectively.

The default arrangement is that the `\textfont` should be a ten-point font, while scripts are typeset at seven points and scripts of scripts are typeset at five points. Like almost anything else in TeX, this can be changed.

TeX also predefines eight of the sixteen allowed families. The predefined families are:

0 Roman `cmr10`, `\rm`; VTeX `mvr10`

1 Math Italics `cmmi10`; VTeX `mvmi10 slant 1000`

2 Math Symbols `cmsy10`; VTeX `mvsy10 slant 1000`

3 Math Extensions `cmex10`; VTeX `mvex10 slant 1000`

4 Italics `cmti10`, `\it`; VTeX `mvu10 slant 250`

5 Slant `cmsl10`, `\sl`; VTeX `mvr10 slant 167`

6 Bold Face `cmbx10`, `\bf`; VTeX `mvbx10`

7 TeleType `cmtt10`, `\tt`; VTeX `mvtt10`

The remaining eight families are left undefined. To define a new family, the user should use the `\textfont`, `\scriptfont`, and `\scriptscriptfont` commands.

SYNTAX ▷ `\textfont`⟨family number⟩ [=] ⟨font command⟩
▷ `\scriptfont`⟨family number⟩ [=] ⟨font command⟩
▷ `\scriptscriptfont`⟨family number⟩ [=] ⟨font command⟩

where ⟨family number⟩ is a number between 0 and 15 and ⟨font command⟩ is a font command defined with `\font`.

EXAMPLE 4.63 ▷ Typing

```
\font\tf=mvtekr scaled 1200 \textfont9=\tf
\font\sf=mvtekr scaled 1000 \scriptfont9=\sf
\font\ssf=mvtekr scaled  800 \scriptscriptfont9=\ssf
```

defines family 9, with the text font at 12 points, scripts at 10 points, and scripts of scripts at 8 points.

The commands `\tf`, `\sf` and `\ssf`, defined above, can be used to switch fonts in text mode. However, they are ignored (or, rather, suspended) when occur in math mode. In

```
$$ \tf\sum_{i=0}^{1992} S_{j_i}\hbox{ (Murphy's formula) }$$
```

`\tf` does not affect the formula; however, it does affect the contents of the `\hbox`:

$$\sum_{i=0}^{1992} S_{j_i} \text{ (Murphy's formula)}$$

Font changes inside math mode are accomplished with the `\fam` command:

SYNTAX ▷ `\fam`⟨family number⟩

EXAMPLE 4.64 ▷ Typing

```
$$\fam9\sum_{i=0}^{1992} S_{j_i}\hbox{ (Murphy's formula) }$$
```

produces

$$\sum_{i=0}^{1992} S_{j_i} \text{ (Murphy's formula)}$$

The reason for separating math- and text- font change commands is that font switches in math mode should consider scripts. Sometimes it may be desirable to design a font switch command to work in both text and math. For instance, TEX defines `\rm` as follows:

```
\font\tenrm=cmr10          \textfont0=\tenrm
\font\sevenrm=cmr7         \scriptfont0=\sevenrm
\font\fivrm=cmr5           \scriptscriptfont0=\fivrm
\def\rm{\fam0\tenrm}
```

where the `\tenrm` part is intended for text mode, while `\fam0` works in math mode.

Font families 4 throught 15 can be freely redefined. However, font families 0, 1, 2 and 3 should not be changed unless they are changed to fonts with layouts identical to `cmr10`, `cmmi10`, `cmsy10` and `cmex10`, respectively. This is because many math character definitions in TEX explicitly refer to these fonts and the character position. For example, the definitions of `\alpha` and and `\forall` are

```
\mathchardef\alpha ="010B
\mathchardef\forall="0238
```

implying that the character α is to be found in the 11th position of family 1 (Math Italics), while the character $\forall$ occupies the 56th position of family 2. The full syntax of `\mathchardef` is

SYNTAX ▷ `\mathchardef`⟨command⟩ [=] = ⟨abcc⟩

where ⟨command⟩ is the name of the symbol being defined, and ⟨abcc⟩ is a hexadecimal number in which a represents the category, b the the ⟨family⟩ number, and cc the symbol position in the font (see §5.2 for the details on hexadecimal notation). When new symbol fonts are added to TeX, it helps to provide mnemonic definitions for the new symbols.

4.21 More Examples

The examples in this section show how you can apply the commands discussed in Chapter 4 to typeset more complicated expressions.

```
$d_ie_i\equiv 1\pmod{\phi (n_i)}$
```

⟹ $d_ie_i \equiv 1 \pmod{\phi(n_i)}$

```
$\psi (p_1^{t_1}),\psi(p_2^{t_2}),\ldots,\psi (p_m^{t_m})$
```

⟹ $\psi(p_1^{t_1}), \psi(p_2^{t_2}), \ldots, \psi(p_m^{t_m})$

```
$2^{170}\not\equiv\big [{2\over 341}\big ]\pmod {341}$
```

⟹ $2^{170} \not\equiv \big[\frac{2}{341}\big] \pmod{341}$

```
$b^N\alpha =(a_na_{n-1}\ldots a_2a_1
 \cdot{\overline{c_1c_2\ldots c_r}})_b$
```

⟹ $b^N\alpha = (a_na_{n-1}\ldots a_2a_1 \cdot \overline{c_1c_2\ldots c_r})_b$

```
$\left|{f(z)-f(z_j)\over z-z_j}-f'(z_j)\right
|<\varepsilon\quad (j=1,2,\ldots,n)$
```

⟹ $\left|\frac{f(z)-f(z_j)}{z-z_j} - f^{\sim}(z_j)\right| < \varepsilon \quad (j = 1, 2, \ldots, n)$

```
$E\varphi=-{h^2\over 2m}{d^2\psi\over dx^2}
                  +{1\over 2}kx^2\varphi$
```

⟹ $E\varphi=-{h^2\over 2m}{d^2\psi\over dx^2}+{1\over 2}kx^2\varphi$

```
$\mu _{i,soln}=\mu _i^*+RT\ln x_i$
```

⟹ $\mu_{i,soln}=\mu_i^*+RT\ln x_i$

```
$P(c_2|c_1)={P(c_1\cap c_2)\over P(c_1)}$
```

⟹ $P(c_2|c_1)={P(c_1\cap c_2)\over P(c_1)}$

```
$Pr(\mu c_2\sigma <X<\mu +2\sigma)=
 \sum_{x=1}^5{9\choose x}
{1\overwithdelims()3}^x{2\overwithdelims()3}^{9-x}$
```

⟹ $Pr(\mu c_2\sigma <X<\mu +2\sigma)=\sum_{x=1}^5\binom{9}{x}\left(\frac{1}{3}\right)^x\left(\frac{2}{3}\right)^{9-x}$

```
$F(x)={1\over \Gamma(\alpha)\beta^\alpha}x^{\alpha -1}
e^{-x/\beta},\qquad 0<x<\infty$
```

⟹ $F(x)={1\over \Gamma(\alpha)\beta^\alpha}x^{\alpha -1}e^{-x/\beta},\qquad 0<x<\infty$

```
$ Pr(R=2k)=
  {2{m-1\choose k-1}{n-1\choose k-1}\over{m+n\choose n}}$
```

⟹ $Pr(R=2k)=\frac{2\binom{m-1}{k-1}\binom{n-1}{k-1}}{\binom{m+n}{n}}$

```
$I_2=\int_{c_2} z^2\,dz=\int _{\sc OA}z^2\,dz
                        +\int _{\sc AB}z^2\,dz$
```

⟹ $I_2=\int_{c_2} z^2\,dz=\int _{OA}z^2\,dz+\int _{AB}z^2\,dz$

```
$F(z+\triangle z)-F(z)=\int _{z_0}^{z+\triangle z}
f(s)\,ds-\int _{z_0}^zf(s)\,ds$
```

⟹ $F(z+\triangle z)-F(z)=\int _{z_0}^{z+\triangle z}f(s)\,ds-\int _{z_0}^zf(s)\,ds$

```
$G(y)=\int _0^{\sqrt y}\int _0^{2\pi}\int _0^\pi{1\over
(2\pi)^{3/2}}e^{-\rho ^2/2}\rho ^2\sin \varphi \,d\varphi
\,d\theta\,d\rho$
```

$\Longrightarrow$ $G(y)=\int _0^{\sqrt y}\int _0^{2\pi}\int _0^\pi{1\over (2\pi)^{3/2}}e^{-\rho ^2/2}\rho ^2\sin \varphi \,d\varphi \,d\theta\,d\rho$

```
$\eqalign{M(t;n)&=E\left[exp\left(t{\Sigma X_i-n\mu\over
\sigma\sqrt n}\right)\right]\cr &=\left[m\left({t\over
\sigma\sqrt n}\right)\right]^n,\qquad -h<{t\over\sigma
\sqrt n}<h}$
```

$$\Longrightarrow \quad \begin{aligned} M(t;n)&=E\left[exp\left(t{\Sigma X_i-n\mu\over \sigma\sqrt n}\right)\right]\\ &=\left[m\left({t\over \sigma\sqrt n}\right)\right]^n,\qquad -h<{t\over\sigma \sqrt n}<h \end{aligned}$$

```
$f(x)=\cases{{m^xe^{-m}\over x!},& $x=0,1,2,\ldots$;\cr
0\hfill,&elsewhere.}$
```

$$\Longrightarrow \quad f(x)=\begin{cases}{m^xe^{-m}\over x!}, & x=0,1,2,\ldots;\\ 0\quad, & \text{elsewhere.}\end{cases}$$

```
$J=\left\vert\matrix{
\partial x_1\over \partial y_1&
\partial x_1\over\partial y_2\cr
\partial x_2\over\partial y_1&
\partial x_2\over \partial y_2}\right\vert
=\left\vert\matrix{1\over 2 & 1\over 2 \cr
1\over 2 & -{1\over 2}}\right\vert
=-{1\over 2}$
```

$$\Longrightarrow \quad J=\left\vert\begin{matrix}\frac{\partial x_1}{\partial y_1} & \frac{\partial x_1}{\partial y_2}\\ \frac{\partial x_2}{\partial y_1} & \frac{\partial x_2}{\partial y_2}\end{matrix}\right\vert=\left\vert\begin{matrix}\frac12 & \frac12\\ \frac12 & -\frac12\end{matrix}\right\vert=-\frac12$$

```
$\bigtriangledown\times F={1\over r^2\sin\varphi}\left|
\matrix{u_r&ru_\varphi&r\sin\varphi u_\theta \cr
\partial\;\over\partial r &\partial\;\over\partial\varphi&
\partial\;\over\partial\theta \cr
F_r&rF_\varphi&r\sin\varphi F_\theta}\right|
```

$$\Longrightarrow \quad \boxed{\nabla \times F = \frac{1}{r^2 \sin\varphi} \begin{vmatrix} u_r & ru_\varphi & r\sin\varphi u_\theta \\ \frac{\partial}{\partial r} & \frac{\partial}{\partial \varphi} & \frac{\partial}{\partial \theta} \\ F_r & rF_\varphi & r\sin\varphi F_\theta \end{vmatrix}}$$

```
$Q=\int\!\!\!\int\!\!\!\!\int _{\cal R}\kappa
\bigtriangledown ^2\tau\, d\tau$
```

$$\Longrightarrow \quad \boxed{Q = \iiint_{\mathcal R} \kappa \bigtriangledown^2 \tau \, d\tau}$$

```
$H_2SeO_3+6I^-+4H^+=Se(s)+2I^-+3H_2O$
```

$$\Longrightarrow \quad \boxed{H_2SeO_3 + 6I^- + 4H^+ = Se(s) + 2I^- + 3H_2O}$$

```
$Zn|Zn^{2+}||Cu^{2+}|Cu$
```

$$\Longrightarrow \quad \boxed{Zn|Zn^{2+}\|Cu^{2+}|Cu}$$

```
$\eqalign{N_2O_3\longrightarrow&N_2O_4+O_2 \cr
&\Updownarrow\cr &2NO_2}$
```

$$\Longrightarrow \quad \boxed{\begin{aligned} N_2O_3 \longrightarrow & N_2O_4 + O_2 \\ & \Updownarrow \\ & 2NO_2 \end{aligned}}$$

```
$\matrix{&A&\displaystyle\buildrel\triangle H_2\over
\longrightarrow \atop T_1&B\cr
\int _{T_2}^{T_1}\overline C_{P,A}\,dT&\Big\downarrow&&
\Big \uparrow&\int _{T_1}^{T_2}\overline C_{P,B}\,dT\cr
&A&\displaystyle\buildrel\triangle H_1\over\longrightarrow
\atop T_2&B}$
```

$$\Longrightarrow \quad \boxed{\begin{matrix} & A & \overset{\triangle H_2}{\underset{T_1}{\longrightarrow}} & B & \\ \int_{T_2}^{T_1} \overline{C}_{P,A}\, dT & \Big\downarrow & & \Big\uparrow & \int_{T_1}^{T_2} \overline{C}_{P,B}\, dT \\ & A & \overset{\triangle H_1}{\underset{T_2}{\longrightarrow}} & B & \end{matrix}}$$

```
$P=\root 3 \of {-q/2+\sqrt{\underline p^3/{27}+q^2/4}}$
```

$$\Longrightarrow \quad \boxed{P=\root 3 \of {-q/2+\sqrt{\underline p^3/{27}+q^2/4}}}$$

```
$nG=\Sigma(n_iG_i^\circ )+
    RT\Sigma(n_i\ln{\hat f_i\over f_i^\circ})$
```

$$\Longrightarrow \quad \boxed{nG=\Sigma(n_iG_i^\circ)+RT\Sigma(n_i\ln{\hat f_i\over f_i^\circ})}$$

```
$G_c(s)=K_c\left (1+{1\over T_{i^s}}\right )(1+T_{d^s})$
```

$$\Longrightarrow \quad \boxed{G_c(s)=K_c\left (1+{1\over T_{i^s}}\right)(1+T_{d^s})}$$

```
$f_c=0.08N_{Re}^{-0.25}+0.01{D\over D_c}^{0.5}$
```

$$\Longrightarrow \quad \boxed{f_c=0.08N_{Re}^{-0.25}+0.01{D\over D_c}^{0.5}}$$

```
${(d_p^\circ )^3\rho _c^2g\over\mu _c^2}=29.0{p^3g_c^3
\overwithdelims () v^3\rho_c^2\mu _cg^4}^{-0.32}{\rho _c
\sigma ^3g_c^3\overwithdelims () \mu _c^4g}^{0.14}$
```

$$\Longrightarrow \quad \boxed{{(d_p^\circ)^3\rho _c^2g\over\mu _c^2}=29.0\left(\frac{p^3g_c^3}{v^3\rho_c^2\mu _cg^4}\right)^{-0.32}\left(\frac{\rho _c\sigma ^3g_c^3}{\mu _c^4g}\right)^{0.14}}$$

```
$G(j\omega)=z{\int_0^\infty e^{-j\omega t}y(t)\,dt \over
\int _0^\infty e^{-j\omega t}xy(t)\,dt}$
```

$$\Longrightarrow \quad \boxed{G(j\omega)=z{\int_0^\infty e^{-j\omega t}y(t)\,dt \over \int _0^\infty e^{-j\omega t}xy(t)\,dt}}$$

```
$\left[{1\over a}\ln{\triangle
   t\over a+b\triangle t}\right]
_{\triangle t_1}^{\triangle t_2}={1\over a}\ln{U_{\circ 1}
\triangle t_2\over U_{\circ 2}\triangle t_1}$
```

$$\Longrightarrow \quad \boxed{\left[{1\over a}\ln{\triangle t\over a+b\triangle t}\right]_{\triangle t_1}^{\triangle t_2}={1\over a}\ln{U_{\circ 1}\triangle t_2\over U_{\circ 2}\triangle t_1}}$$

Chapter 5

Variables Dimensions Glue

The great majority of documents can be typeset by using only the high-level commands. However, an advanced user of the package should be familiar with the internal low-level structures used in the language in order to truly understand TEX and be able to use all its power. These internal structures are introduced in this and subsequent chapters.

Knowing the low-level structures and operations you can:

- use abbreviation (simple text macros)
- implement new commands and stylesheets
- do conditional processing of the text
- perform file operation from TEX
- interact with the user
- compute references for indices and tables of contents

5.1 Variables: General Information

The variables of TEX come in several types, as do the variables of other programming languages. There are, among others:

- *count* variables. They hold integer parameters, such as page numbers, line counts, footnote numbering, or day of the month. The count variables are described in §5.2.

- *dimen*sion variables. They hold dimensional parameters or distances, such as paragraph indentation and page dimensions (see §5.3).

The other types of variables are less traditional. For instance,

- *box* variables hold entire rectangles of typeset text. They permit us to repeat a portion of text without typesetting it each time (see §8.4).

- *glue* or *skip* variables specify varying (stretchable) dimensions. These are similar to dimen variables (see §5.7).

Many of the commands introduced in the previous chapters can alternatively be viewed as variables. For example, `\hsize`, introduced at the beginning of the Chapter 3, may be viewed as a command that sets the horizontal size of the page to a particular amount, as in

```
\hsize 5in
```

Or one may view it as an assignment of the value of `5in` to the `\hsize` (dimen) variable, as in

```
\hsize=5in
```

There is no contradiction between these points of view since there are also two ways to view TEX:

- as a command-driven typesetting system, where it would be natural to take `\hsize 5in` as a command.
- as a programming language for typesetting, where it would be natural to take `\hsize` as a variable and `\hsize=5in` as an assignment.

The optionality of the equal sign in the assignment statements allows these two points of view to coexist.

Other variables already introduced as commands include:

- count variables `\pageno`, `\enumcount`, and `\hangafter`
- dimen variables `\hoffset`, `\parindent`, and `\hangindent`
- glue variables `\parskip` and `\baselineskip`
- *shape* variable `\parshape` (see §3.7.)

The assignment syntax, including the optional equal sign, is a feature that is common to all types of variables. Other features that are common to most types include:

- New variables can be declared. Declaration commands `\newcount`, `\newdimen`, and `\newtoks` are used to create new count, dimen, and token string variables (see §6.18.)
- The value of a variable can be inserted in the text by using `\number`, `\the`, and other commands.
- The contents of the variables can be displayed on the terminal in interactive mode by using the `\showthe` command.
- Arithmetic operations can be performed on arithmetic types of variables, i.e., count, dimen, and glue. For example, `\multiply\var by 2` doubles the value of a variable `\var`.
- Variables are normally local to the groups in which they are used.

This chapter formally introduces and discusses the simpler types of variables, such as count, dimen and glue. More advanced concepts, such as box variables, are discussed later in the book.

5.2 Numbers and Count Variables

TOPICS ▷ ⟨number⟩, ⟨count⟩, `\newcount`, `\advance`, `\divide`, `\multiply`
▷ `\number`, `\upr`, `\romannumeral`, `\the`, `\showthe`, `\count`, `\global`

The simplest type of TEX variable is the count, or integer, variable. These can hold (positive or negative) integer numbers.

There are four ways to enter integer numbers:

- as decimal numbers: `10`, `1987`, `-783`
- as octal numbers. An octal number begins with a single closing quote, followed by octal digits $0, 1, 2, \ldots, 7$. For example, `'10` is the same as decimal number 8; `-'772` and `'477` are valid octal numbers.
- as hexadecimal (*hex*) numbers. They begin with a double quote, followed by hexadecimal digits $0, 1, 2, \ldots, 9, A, B, \ldots, F$. For example, `"10` is the same as decimal 16; `-"CC` and `"1ABC` are valid hex numbers.
- as the codes representing ASCII characters. These contain an opening quote, followed by the character. For example, `` `a `` defines the decimal number 97 since the lower-case 'a' is the 97-th character in ASCII. You can also write `` `\a ``; this too defines 97. To enter the values of low ASCII characters (like Ctrl-P or ASCII 16), you may also use the `^^P` abbreviation (this is the same as decimal 16). Finally, you can also type `^^10`; since `"10` is the same as `16`.

The count variables include, for instance:

- `\pageno`, the current page number
- `\year`, the current Christian Year (Common Era)
- `\month`, the current month
- `\day`, the current day
- `\footcount`, the current tag for numbered footnotes

The value of a count variable can be inserted in the document by preceding its name with the `\number` or `\the` prefix commands.

EXAMPLE 5.1 ▷ Typing

```
This page of the book is page \the\pageno.
It was printed on \number\month/\number\day/\number\year.
The next footnote will be numbered \number\footcount.
```

produces

This page of the book is page 137. It was printed on 8/2/1992. The next footnote will be numbered 1.

The `\number` and `\the` commands are interchangeable for count variables, but `\the` is more general, since it also works with other types of variables. The `\romannumeral` prefix typesets roman lower case; standard TeX makes printing upper case roman numerals a bit tricky, since

```
\uppercase{\romannumeral⟨number⟩}
```

does not work. The `\upr` command, described in §6.8 does the trick. All prefix commands may also be used with integer constants.

EXAMPLE 5.2 ▷ Typing

```
The French Revolution started in the year
\romannumeral 1789\ Anno Domini.   By the year
\upr{\romannumeral 4} of the New Calendar
guillotines were erected and the real fun began.
```

produces

The French Revolution started in the year mdcclxxxix Anno Domini. By the year IV of the New Calendar guillotines were erected and the real fun began.

The `\showthe` command displays the value of a count variable on the screen.

EXAMPLE 5.3 ▷ Typing

```
\showthe\year
```

results in

```
> 1992.
<*> \showthe\year
```

printed on the terminal (and to the log file). TeX then displays the ?-prompt (see §1.6) and pauses for your input. You can resume typesetting by pressing the carriage-return key.

To set the value of a count variable, use the assignment statement.

SYNTAX ▷ ⟨count variable⟩ [=] ⟨number⟩

where ⟨number⟩ refers to an integer number, some other count variable, or any expression that produces an integer number. If the right hand side of the assignment does not contain a number, you will get the error message

```
! Missing number, treated as zero.
```

and the ⟨count variable⟩ will be set to zero. The equal sign (=) is optional in assignments.

You can declare new count variables with the `\newcount` command.

SYNTAX ▷ `\newcount` ⟨count variable⟩

For example, by typing `\newcount\savecount` you create the new count variable `\savecount`. It will be initialized to 0. New (user-defined) count variables can be treated in the same way as the pre-defined variables. In particular, you can type

```
\savecount=88
\savecount=\pageno
```

to set `\savecount` to number 88 or the current page number, respectively.

You can define variables, for instance, to

- keep track of footnote numbering. The same footnote may be referred to from several places in the document.
- keep track of page and section numbering to automate references within the document.
- do simple computation during typesetting (see below).

EXAMPLE 5.4 ▷ Suppose that you are writing a fairly large manuscript that contains page references. You might have to typeset text similar to *"Dementia leprosis nowadays is easily treated by placebo if diagnosed in time; see page 743 for the symptoms"*. The problem is that you would not know the exact page number until the document has actually been typeset. A solution is to save the `\pageno` of the page that discusses the symptoms:

```
\newcount\dementiapage \dementiapage=\pageno
```

and, instead of typing

```
... see page 743 for the symptoms
```

type

```
... see page \the\dementiapage\ for the symptoms
```

The correct reference will be computed automatically.

➡ With this technique you may refer only to previous pages of the document. An improved scheme that allows forward references is presented in §7.3.

Arithmetic operations (addition, subtraction, multiplication, and division) can be performed on the count variables.

SYNTAX ▷ `\advance` ⟨count variable⟩ `by` ⟨number⟩

adds ⟨number⟩ to ⟨count variable⟩. ⟨number⟩ may be preceded by the minus sign '-' for subtraction.

SYNTAX ▷ `\multiply` ⟨count variable⟩ `by` ⟨number⟩

multiplies ⟨count variable⟩ by ⟨number⟩.

SYNTAX ▷ `\divide` ⟨count variable⟩ `by` ⟨number⟩

divides ⟨count variable⟩ by ⟨number⟩.

EXAMPLE 5.5 ▷ Typing

```
\newcount\nextyear \nextyear=\year \advance\nextyear by 1
Dear customer, the rest of your order will be promptly
shipped in the last quarter of \the\nextyear.
```

produces

> Dear customer, the rest of your order will be promptly shipped in the last quarter of 1993.

in all letters mailed in 1992.

Variables are always local to the groups in which they are declared. If you enter

```
\newcount\myvar \myvar=5 {\myvar=7}
```

`\myvar` will be set to 5. Use the `\global` prefix command to permanently change a variable from inside a group. For instance,

```
\newcount\myvar \myvar=5 {\global\myvar=7}
```

sets `\myvar` to 7.

The total number of count variables is 256. The count variables may be viewed as drawers that hold integer values. The drawers are numbered 0 through 255 and can be accessed explicitly by specifying the variable number. For instance, `\count74` accesses the seventy fourth ⟨count⟩ variable. The `\newcount\newvar` command simply assigns the label `\newvar` to one of the available count variables. You may find the drawer number of a particular count variable with the `\show` command.

EXAMPLE 5.6 ▷ Typing

```
*\newcount\newvar \newvar=\count25  \show\newvar
```

causes TeX to print

```
> \newvar=\count25.
<*> \show\newvar

?
```

5.3 Dimensions and Dimensional Variables

TOPICS ▷ ⟨dimen⟩, `\newdimen`, `\dimen`
▷ `pt, pc, in, bp, cm, mm, dd, cc, sp, em, ex, mu, true`

Dimensional constants and variables specify page measurements, spacing, paragraph indentation, and other measurable items. There are many ways to specify dimensions. You can use both English and metric measurements as well as the traditional units used by professional typesetters both in the U.S. and in Europe. Dimensions appear in many of the TeX commands and are specified according to the following general rule:

SYNTAX ▷ ⟨number⟩ ⟨unit of measure⟩,

where ⟨number⟩ may be positive or negative; it may contain a decimal dot (or comma); it may also be entered as a decimal, octal, or hex number. ⟨unit of measure⟩ is one of the two-letter abbreviations of dimensional units listed in the following table:

Dimensions in TeX			
Abbr.	Name	Used in ...	Conversion
`pt`	point	U.S. typesetting	
`pc`	pica	U.S. typesetting	1 `pc` = 12 `pt`
`in`	inch	English system	1 `in` = 72.27 `pt`
`bp`	big point	U.S. typesetting	72 `bp` = 1 `in`
`cm`	centimeter	Metric System	2.54 `cm` = 1 `in`
`mm`	millimeter	Metric System	25.4 `mm` = 1 `in`
`dd`	didot point	European typesetting	1157 `dd` = 1238 `pt`
`cc`	cicero	European typesetting	1 `cc` = 12 `dd`
`sp`	scaled point	TeX text unit	655536 `sp` = 1 `pt`
`em`	m-width	TeX text unit	font dependent
`ex`	x-height	TeX text unit	font dependent
`ex`	mu	TeX math unit	see §4.14

The following are valid dimensions:

```
0.5 in,   +7.35 cm,   -11,2cc,   -10241024sp,   5.55 mm,   "CC cm.
```

Observe that in the last example the space is needed.

Dimensional specifications must always contain a ⟨unit of measure⟩. If it is missing, as in

```
\parindent=7 Here it goes...
```

you will get the error message

```
*\parindent=7 He
! Illegal unit of measure (pt inserted).
...
```

➡ The dimensional units listed above correspond to the dimensions that will actually appear on paper, provided that the `\magnification` of the document is 1000. If `\magnification` is not 1000, the dimensions are scaled by the magnification factor. For instance, with magnification 1500, `\hskip2in` will actually span 3 inches. When this effect is undesirable, use the `true` suffix: specifying `\hskip2truein` (or `\hskip 2 true in`) will always result in a 2-inch skip.

In this book the dimensions are mostly specified in points (`pt`) since that is the established typesetting unit; also using points allows most dimensions to be expressed as comfortable small integer numbers.

The command `\the` is used to insert the value of a dimensional variable into the document; `\showthe` prints its value on the screen.

EXAMPLE 5.7 ▷ `\hsize` is a dimensional variable that holds the horizontal size of the printable part of the page. Typing

```
This page has horizontal size \the\hsize.
```

produces

This page has horizontal size 397.48499pt.

The dimensional variables include:

- `\vsize`, the vertical size of the page
- `\parindent`, the amount of indentation in the beginning of each paragraph
- `\hangindent`, the amount of hanging indentation

New dimensional variables can be created by using `\newdimen`. The value of a dimensional variable can be changed by using `\advance`, `\multiply`, or `\divide`.

EXAMPLE 5.8 ▷ Typing

```
\newdimen\dimenvar
\dimenvar=3.5in
\advance\dimenvar by 5cm
\multiply\dimenvar by 2
\divide\dimenvar by 3
```

sets `\dimenvar` to $(3.5in + 5cm) * 2/3$.

The `em` and `ex` dimensions are not absolute, but depend on the current font. In each font, `em` value is the width of the em-dash or the size of `\quad`, while `ex` is equal to the x-height of the font. These relative dimensions should be used in the cases when a portion of text that contains precise positioning may appear in different fonts. For example, the `\TeX` logo command uses precise kerning to position the characters (see §8.5).

Internally, all dimensions are converted to scaled points. The largest dimension that can be specified by the user is 2^{30}`sp`, which is about 19 feet.

As with the ⟨count⟩ variables, there are 256 ⟨dimen⟩ variables, numbered 0 through 255. They can be accessed explicitly. For instance, `\dimen23` accesses the twenty-third ⟨dimen⟩ variable. The `\newdimen\newvar` command simply assigns one of the available ⟨dimen⟩ variables to `\newvar`. Given a ⟨dimen⟩ variable, you may find its internal number with the `\show` command. The ⟨dimen⟩ variables are also local to their groups; the `\global` command is used for global changes (see §5.2).

5.4 Inter-word Spacing. Glue

TOPICS ▷ shrink, stretch, `plus`, `minus`

All dimensional specifications are exact. However, in many cases, specifying dimensions exactly is impossible. This is the case, for instance, with interword spacing.

Recall that in formatting a paragraph, TeX tries to set the interword spacing within certain limits and keep it as close to the ideal spacing as possible. The interword spacing is, in fact, defined by three dimensions:

- *interword space*, the ideal spacing between words
- *interword stretch*, the maximum extra amount of space which can be added to the interword space

- *interword shrink*, the maximum amount of space which can be taken from the interword space

This triad of dimensional parameters is font dependent, since spacing rules are somewhat different in different fonts. For example, in the 10-point roman font `cmr10`, the interword space is set to 3.33 points, the interword stretch is 1.67 points, and the interword shrink is 1.11 points. This means that the ideal interword spacing for `cmr10` is 3.33 points, the maximum allowed spacing is 5.00 points (since 5.00 = 3.33 + 1.67), and the minimum allowed spacing is 2.22 points. Formally, one can express this by saying that the interword spacing in `cmr10` is

```
3.33pt plus 1.67pt minus 1.11pt
```

The interword spacing in other fonts is different. For example,

- in `cmbx10` it is `3.83pt plus 1.92pt minus 1.28pt`. All dimensions are a bit bigger than in `cmr10` to compensate for the **boldness**.
- in `cmsl10` it is `3.33pt plus 1.67pt minus 1.11pt`, just as in roman. The reason for the equality is that these two fonts are essentially the same, except for the *slant*.
- in `cmtt10` it is `5.25pt plus 0pt minus 0pt`. The `typewriter font` is not proportional and neither is its spacing—it cannot be stretched or shrunk.
- in `cmmi10` it is `0pt plus 0pt minus 0pt`. *Math Italics* are designed for equations, not for text, and do not have spacing.

The interword parameters are used by commands that format lines and paragraphs of text, including the `\line` command (see §3.2). Its inner workings can be described by the following physical analogy:

1. Cut out a small metal plate (like the ones used by linotype) for each word in the line.
2. Make a drawer that is exactly as wide as the line of text will be.
3. Drop the plates in the drawer in the right order.
4. Prepare a liquid glue with the property that when left alone, each drop of glue occupies exactly 3.33 points of space (if the current font is roman; different glues would be needed for other fonts), but each drop of glue can be stretched another 1.67 points or be compressed (shrunk) 1.11 points (so the glue is a bit like a spring—it can stretch and shrink.).
5. Put a drop of glue between the word plates.
6. Shake the drawer and wait until the glue dries up (actually, until TeX finishes its job).

If the interword glue has to be stretched beyond its *stretch*, the glue will break and the line will become underfull. If it has to be shrunk more than physically possible, the drawer will break, words plates will stick out, and the line will become overfull. If neither of these happens, the line is typeset properly.

➡ The metal plates are called *boxes;* they are described in Chapter 7.

In light of this analogy, the triad of the interword parameters is called *interword glue*. More generally, any triad of dimensions that specify normal dimensions, stretch and shrink is called *glue*.

The interword glue is not the only kind of glue used in paragraph formatting. The intersentence glue is used to determine the space between sentences. In `cmr10`, the intersentence glue is

```
4.44pt plus 1.67pt minus 1.11pt
```

5.5 Infinite Glues. Filling

TOPICS ▷ `fil`, `fill`, `\hfil`, `\hfill`, `\hss`, `\hfilneg`

You can make your own kinds of glue and use them, for instance, in the `\hskip` command (horizontal skip, see §3.5).

EXAMPLE 5.9 ▷ Typing

```
\line{This is a sample line}
```

produces a badly formatted line

This	is	a	sample	line

and an underfull warning. You can format the line by inserting a stretchable glue anywhere in the line. For instance, typing

```
\line{This is \hskip 0pt plus 10in minus
0pt a sample line}
```

produces

This is	a sample line

The '`0pt plus 10in minus 0pt`' glue can expand up to 10 inches, so it will be sufficient to cover all the extra space. Putting this glue at the end of the sentence will produce a left-justified line, just as if you had used the `\leftline` command;

putting it at the beginning of a sentence will produce a right-justified line, just like `\rightline`. In addition, you can put it both at the beginning and at the end, thus centering the line.

Instead of using large dimensions, you can plainly state that the glue can be infinitely stretched. To do it, use the special dimensional unit `fil`. In the above example, one could have used the glue

```
0pt plus 1fil minus 0pt
```

to achieve the same effects.

The `\hfil` command is the abbreviation (not a macro but hard-wired into the system for faster processing) for

```
\hskip 0pt plus 1fil minus 0pt
```

It can be used to left-justify, right-justify, or center text, or produce unusual formats, as in the following example:

EXAMPLE 5.10 ▷ Typing

```
\line{\hfil Mary had a \hfil little lamb, \hfil}
\line{\hfil Whose fleece was \hfil white as snow. \hfil}
\line{\hfil And everywhere \hfil that Mary went. \hfil}
\line{\hfil The lamb was \hfil sure to go. \hfil}
```

produces

Mary had a	little lamb,
Whose fleece was	white as snow.
And everywhere	that Mary went.
The lamb was	sure to go.

In this example, the leftover space in each line is evenly distributed between `\hfil`'s.

The `fill` dimensional unit also represents infinite dimensions, but it is "more infinite" than `fil`. The `\hfill` command is the abbreviation for

```
\hskip 0pt plus 1fill minus 0pt
```

If both `\hfil` and `\hfill` are present, `\hfill` overrides the action of `\hfil`.

EXAMPLE 5.11 ▷ Typing

```
\line{\hfil Mary had a little lamb. \hfill}
```

produces

Mary had a little lamb.

The `\hfill` command can be used to change the behavior of commands that use `fil` for alignment. For instance, the `\pmatrix` command (see §4.12) normally centers the entries of the matrix. You can right-align them by inserting an `\hfill` before each entry.

EXAMPLE 5.12 ▷ Typing

```
$$\pmatrix{1 & 17878 \cr 22233 & 6 },
  \pmatrix{\hfill     1 & \hfill 17878 \cr
           \hfill 22233 & \hfill     6 } $$
```

produces

$$\begin{pmatrix} 1 & 17878 \\ 22233 & 6 \end{pmatrix}, \begin{pmatrix} \hfill 1 & \hfill 17878 \\ \hfill 22233 & \hfill 6 \end{pmatrix}$$

➡ It is best not to use `\hfill` where `\hfil` would be sufficient.

In addition to infinitely stretchable glues, there are also glues that are infinitely shrinkable. For example, the `\hss` (horizontal shrinkable and stretchable) glue is defined as

```
\hskip 0pt plus 1fil minus 1fil
```

EXAMPLE 5.13 ▷ Typing any one of

```
\centerline{  This is a centered line.       }
\line{ \hfil  This is a centered line. \hfil }
\line{ \hfill This is a centered line. \hfill}
\line{ \hss   This is a centered line. \hss  }
```

produces a centered line

This is a centered line.

The difference will show up if the line to be centered is too long to fit, that is, longer than `\hsize`. The `\hfil` and `\hfill` glues cannot be shrunk; an attempt to typeset a long line with either of them will cause an overfull error. The `\hss` glue will shrink; as a result, the beginning and the end of the line will occur on the margins, but no error will be generated. Since it may be acceptable to allow longer titles to intrude into the margins, the `\centerline` command uses the `\hss` glue.

EXAMPLE 5.14 ▷ Typing

```
\line{\hss This is a very, very, very,
very, very, very, very, very, very,
very, very, very, very long title \hss}
```

produces

This is a very, very, very, very, very very, very, very, very, very, very, very, very, very, very long title

A similar example that used `\hfil` or `\hfill` instead of `\hss` would be typeset (with an overfull error message) as

This is a very, very, very, very, very, very, very, very, very, very, very, very, very, very, very, very long title

Notice that the second example generates an overfull error message, while the first does not.

➡ The `\leftline` and `\rightline` commands also use `\hss`.

The last predefined infinite glue is the glue generated by `\hfilneg`. It is in effect an abbreviation for

```
\hskip 0pt plus -1fil
```

If an `\hfilneg` glue is placed next to an `\hfil` glue, it cancels the effects of `\hfil`. This feature can be used to suppress or override alignment generated by commands that use the `\hfil` glue.

5.6 Vertical Glues

TOPICS ▷ `\vfil`, `\vfill`, `\vss`, `\vfilneg`

Glues are also used in the vertical mode to typeset pages from lines. The physical analogy given in §5.4 is still applicable, but in the vertical mode the metal plates represent entire lines, already typeset with the horizontal glues, and the drawer has the shape of a page. Vertical glue is added between lines.

The `\baselineskip` and `\parskip` commands (see §3.12) define the vertical glue that is used between lines in a paragraph and the extra amount of vertical glue that is added between separate paragraphs. By default, the `\baselineskip` glue is set to

```
12.0pt plus 0.0pt minus 0.0pt
```

(or simply to `12pt`) and the `\parskip` glue is

```
0.0pt plus 1.0pt minus 0.0pt
```

This means that the interline spacing is not stretchable, but up to a point of extra space may be inserted between paragraphs.

The `\vfil`, `\vss`, `\vfill`, and `\vfilneg` commands are defined and operate like their horizontal counterparts.

EXAMPLE 5.15 ▷ The `\page` command (see §3.14) abbreviates `\vfill\eject`. Without a `\vfill`, `\eject` causes the extra vertical space to be equally divided between the lines of the page being ejected; compare this to the `\line` command in the horizontal mode. The `\vfill` command collects all the extra space at the bottom of the page, pushing the text up.

5.7 Glue Variables

TOPICS ▷ `\newskip`, `\smallskipamount`, `\medskipamount`
▷ `\bigskipamount`, `\newglue`

Glue variables hold amounts of horizontal and vertical glue. The following syntax is used to specify an amount of glue:

SYNTAX ▷ ⟨dimen⟩ [`plus` ⟨dimen⟩ [`minus` ⟨dimen⟩]]

where ⟨dimen⟩ may use all the standard dimensional units (listed in §5.3) as well as `fil` and `fill`.

The following are valid glue specifications: `10pt`, `-3in`, `1fil`, `0pt plus 1in`, `2pt plus 3pt minus 4pt`. However, `3pt minus 2pt` is not a valid glue, since the `plus` term must be specified first. Use instead `3pt plus 0pt minus 2pt`.

The glue variables include

- `\baselineskip`, the interline glue
- `\parskip`, the extra inter-paragraph glue
- `\leftskip`, the extra left margin glue
- `\smallskipamount`, the glue used by the `\smallskip` command

Like the count and dimen variables, glue variables can be inserted in the document or displayed on the screen with the `\the` and `\showthe` commands. Use the assignment statement to change the contents of a glue variable.

There are 256 ⟨glue⟩ variables, numbered from 0 to 255. They can be accessed by number with the `\skip` command. For example, `\skip27` refers to the twenty-seventh glue variable. The `\newskip\myskip` (or `\newglue\myskip`) command allocates one of the 256 glue variables to `\myskip`. The `\show` command can be used to find the number of a particular ⟨glue⟩ variable. ⟨glue⟩ variables are local to their groups; the `\global` command should be used for global changes (see §5.2).

The `\advance`, `\divide`, and `\multiply` commands can be used with glue variables.

EXAMPLE 5.16 ▷ Typing

```
\newskip \myskip
\myskip = 10pt plus 1pt minus 2pt
\advance\myskip by 23pt plus 2pt minus 1fil
```

creates a glue variable `\myskip` and sets it to

```
33pt plus 3pt minus 1fil
```

Notice that the infinite dimensional units like `fil` are not added to the finite dimensions but instead supersede them.

The `\bigskip`, `\medskip` and `\smallskip` commands described in §3.13 generate vertical skips by `\bigskipamount`, `\medskipamount`, and `\smallskipamount`, respectively. These are glue variables, set by default as follows:

```
\smallskipamount =  3.0pt plus 1.0pt minus 1.0pt
\medskipamount   =  6.0pt plus 2.0pt minus 2.0pt
\bigskipamount   = 12.0pt plus 4.0pt minus 4.0pt
```

By changing the values of these variables you change the amount of space that the skip commands generate. For example, inserting

```
\medskipamount = 8pt
```

at the beginning of the source will make all spaces generated by `\medskip` and `\medbreak` in the entire document to be exactly 8 points high.

Some other glue variables of interest are:

- `\topskip`, the extra space to be omitted at the top of each page. By default, it is set to 0pt. Setting `\topskip` to a positive value allows, for instance, typesetting on letterheads.
- `\belowdisplayskip`, the skip used after a display math equation (default: 12.0pt plus 3.0pt minus 9.0pt).

- `\abovedisplayskip`, the skip used before a display math equation (default: 12.0pt plus 3.0pt minus 9.0pt).
- `\belowdisplayshortskip`, the skip used after a display math equation, if the previous line is short (default: 7.0pt plus 3.0pt minus 4.0pt).
- `\abovedisplayshortskip`, the skip used before a display math equation, if the previous line is short (default: 0.0pt plus 3.0pt).

Chapter

6

Macros

The macro facility allows you to abbreviate frequently used character and command sequences. Instead of retyping the same sequences many times, you can define macro commands and simply enter their names. The majority of commands discussed in this book are in fact pre-defined macros; familiarity with the macro facility allows you to define your own commands, change the existing ones, and completely customize TeX.

6.1 Macros as Abbreviations

TOPICS ▷ `\def`, `\show`

The simplest use of the macro facility is to create abbreviations for frequently used phrases.

SYNTAX ▷ `\def` ⟨macro name⟩ `{` ⟨macro text⟩ `}`

where ⟨macro name⟩ is the name of the macro being defined and ⟨macro text⟩ is the text that the macro generates every time it is invoked. ⟨macro name⟩ may contain only letters.

EXAMPLE 6.1 ▷ By typing

```
\def\EEF{Exotic and Edible Foods}
```

you define the new command `\EEF`. Every time you enter the `\EEF` command, the phrase "Exotic and Edible Foods" will be included in the document. For instance, typing

```
Radical changes in nutrition--this is the motto of \EEF.
```

produces

Radical changes in nutrition—this is the motto of Exotic and Edible Foods.

without the necessity to spell out "Exotic and Edible Foods."

Macro abbreviation can be used, for instance, for

- personal names, as in `\def\me{Iohannes Gutenberg}`
- common phrases, as in `\def\sy{Sincerely Yours}`
- long technical terms, as in `\def\DL{Dementia Leprosis}`

TeX does not supply a space after the macro expansion, since in some circumstances this space is not needed. Thus, it would be wrong to type

```
\EEF is a large and prosperous company.
```

since it would appear as

Exotic and Edible Foodsis a large and prosperous company.

Instead, type either

```
\EEF\ is a large and prosperous company.
```

or

```
{\EEF} is a large and prosperous company.
```

➡ It also would be wrong to include a space in the definition of the macro. Indeed, if you define

```
\def\EEF{Exotic and Edible Foods\␣}
```

the space will appear in all cases, including ones where it should not appear (before punctuation signs, for instance). A different way of handling spaces after text macros is discussed in §6.3.

The ⟨macro text⟩ may also contain commands, as well as names of other user macros.

EXAMPLE 6.2 ▷ The following macro definitions are valid:

```
\def\EEF{{\bf Exotic and Edible Foods}}
\def\Remark{\medskip {\it Remark: }}
\def\TM{({\Smallcaps tm})}
```

➡ Double braces are needed in the definitions to limit the action of the font change commands. When macro expansion takes place, the outside braces disappear. Therefore, if only one pair of braces were present, a font command would be inserted in the body of text without surrounding braces and would affect all subsequent text.

EXAMPLE 6.3 ▷ If you enter the definitions from the previous example and type

```
\Remark Digestron\TM\ is a revolutionary product by \EEF.
```

you get

Remark: Digestron(TM) is a revolutionary product by **Exotic and Edible Foods**.

➡ The trademark sign is frequently typeset as a superscript. This can be done with the `\raise` command (see §8.5).

EXAMPLE 6.4 ▷ Macros may also be used to hold a company's logo or a trademark. Fine effects can be achieved by using symbol fonts and/or kerning commands (see §8.5). For instance, typing:

```
\def\EEFlogo{$\cal E\!\cal E\!\cal F$}
```

defines $\cal E\!\cal E\!\cal F$. The definition of the `\TeX` logo uses kerning; it can be found in §8.5.

The `\show` command shows the definition of the macro.

SYNTAX ▷ `\show` ⟨macro name⟩

EXAMPLE 6.5 ▷ Typing

```
\def\EEF{{\bf Exotic and Edible Foods}}        \show\EEF
```

echoes the following to the screen:

```
> \EEF=macro:
->{\bf Exotic and Edible Foods}.
<*> \show\EEF

?
```

After listing the contents of the macro, TeX will go into the interactive mode and display the ?-prompt. Press carriage return to continue.

Macros that are used frequently are often placed in include files or *stylesheets*. For example, the employees of $\cal EEF$ might share a common file `EEF.STY` that would contain frequently used abbreviations such as `\EEF`, `\EEFlogo`, and `\TM`. Every TeX source file in $\cal EEF$ would include the command

```
\input EEF.STY
```

to load the shared macros.

Other more powerful forms of macros are discussed in the following sections.

6.2 Macros with Arguments

TOPICS ▷ `\def`

Macros can be optionally supplied with arguments.

SYNTAX ▷ \def ⟨macro name⟩ [#1[#2...#9]...] { ⟨macro text⟩ },

where ⟨macro name⟩ and ⟨macro text⟩ are as specified in the previous section; **#1** through **#9** specify optional macro arguments; at most nine are allowed.

Each invocation of a macro with arguments should be followed by a number of tokens (single characters) or groups (strings in braces) equal to the number of arguments in the definition of the macro. These tokens or groups are substituted in place of the macro arguments.

EXAMPLE 6.6 ▷ Typing

```
\def\boldface#1{{\bf #1}}
```

defines the command `\boldface` which invokes the **boldface** font. Unlike the `\bf` command, `\boldface` requires its argument to be specified explicitly; if the argument is longer than one character, it must be surrounded by braces. You can use `\boldface`, for instance, by typing

```
This is a \boldface{bold} word.
```

which produces

> This is a **bold** word.

EXAMPLE 6.7 ▷ Typing

```
\def\loan#1#2#3{{#1} lent {#2} \$#3. {#2} owes {#1} \$#3.}
\loan{Jack}{Jim}{10} \loan{Mary}{Beth}{12}
```

results in

> Jack lent Jim $10. Jim owes Jack $10. Mary lent Beth $12. Beth owes Mary $12.

EXAMPLE 6.8 ▷ The definition

```
\def\title#1{\bigbreak\centerline
 {\underbar{\bf #1}}\nobreak\medskip}
```

can be used to format section headers. Typing, for instance,

```
\title{Digestron\TM\ and the future of eating}
```

will produce a centered, bold and underlined section header, as well as a `\bigbreak` before and a `\medskip` after it:

> **Digestron(TM) and the future of eating**

The `\nobreak` command also ensures that a page break will not happen immediately after the section header.

The form of `\def` introduced in this section requires that the macro substitution arguments be surrounded by braces. However, a more general form of `\def`, discussed in the next section, allows one to specify other ways of parsing the arguments.

6.3 Macros with Delimited Arguments

TOPICS ▷ `\def`

The most general form of the macro definition command `\def` allows the use of almost any text characters to delimit the arguments of macros. These delimiting characters are put in the ⟨argument list⟩; when the macro definition is invoked, they guide TEX in finding the arguments.

SYNTAX ▷ `\def` ⟨macro name⟩ ⟨argument list⟩ `{` ⟨macro text⟩ `}`

where ⟨argument list⟩ is the list of arguments #1, #2,...,#9 optionally separated by delimiting characters and ⟨macro text⟩ contains the substitution text. The following rules govern the contents of ⟨argument list⟩ and ⟨macro text⟩:

0. The ⟨argument list⟩ may not contain braces (a brace indicates the beginning of the ⟨macro text⟩).
1. In the ⟨macro text⟩, braces must be matched.
2. In the ⟨argument list⟩, macro symbols #1 through #9 should appear in order. Thus #2 should always appear after the appearance of #1 and can be specified only once, #3 may appear only after #2 and again at most once, and so on.
3. Other characters may appear in the ⟨argument list⟩; they act as delimiters in separating the arguments during the macro expansion.
4. In the ⟨macro text⟩, # may be followed only by one of the digits 1 through 9, in which case an appropriate macro argument will be substituted, or by another #, in which case TEX sees a single character # and inserts it into the expanded text.
5. The ⟨argument list⟩ may start with delimiting characters. In this case, they in effect become part of the command name (this technique can be used to resolve the problem of the after-macro space raised in §6.1).
6. When macro is invoked, its arguments may not exceed a paragraph (see §6.7 for the rationale behind this restriction and a way to overcome it).
7. The macro definition applies only in the current group; after the group is closed, the macro is forgotten (see §6.7, `\global`).

EXAMPLE 6.9 ▷ The `\boldsentence` macro, defined as

```
\def\boldsentence#1.{{\bf #1}.}
```

converts to the bold font exactly one sentence. The period that appears after `#1` acts as a delimiter; all text following `\boldsentence` before the period is viewed as the argument to `\boldsentence` and, therefore, is typeset in boldface. Typing

```
\def\boldsentence#1.{{\bf #1}.}
The next sentence will be bold. \boldsentence This is
a bold sentence. And this one is not.
```

produces

The next sentence will be bold. **This is a bold sentence**. And this one is not.

➡ TEX is not clairvoyant; if the sentence contains any period, only the text preceding the period will become bold. For instance, typing

```
Mr.~User thought that the next sentence would be bold.
\boldsentence Mr.~User was wrong.
```

produces

Mr. User thought that the next sentence would be bold. **Mr**. User was wrong.

➡ Delimiting characters found in the replacement text are not copied into the macro expansion. This is why the definition of `\boldsentence` contains an extra period in the expansion text and should not be entered as

```
\def\boldsentence#1.{{\bf #1}}
```

Similarly, you can use `\par` instead of '.' to create a `\boldparagraph` macro:

```
\def\boldparagraph#1\par{{\bf #1\par}}
```

This macro will typeset the entire next paragraph in the bold font. The paragraph can be delimited by either an explicit `\par` command or by an empty line.

The `\proclaim` macro command, pre-defined in TEX, operates on an entire paragraph. Its definition is essentially

```
\def\proclaim #1. #2\par{\medbreak
{\bf#1.\enspace}{\sl#2}\par\medbreak}
```

➡ The actual definition also makes it an `\outer` macro (see below).

The first sentence that follows `\proclaim` is typeset in bold; an `\enspace` is placed after it; the rest of the paragraph is slanted. Also, `\proclaim` generates a `\medbreak` before and after the paragraph.

EXAMPLE 6.10 ▷ Typing

```
\proclaim Remark 7. Once you start using them,
 macros are rather easy.
```

produces

> **Remark 7.** *Once you start using them, macros are rather easy.*

Delimited and undelimited arguments can be mixed.

EXAMPLE 6.11 ▷ The `\fontsentence` macro generalizes the previously defined `\boldsentence`:

```
\def\fontsentence#1#2.{{#1#2}.}
```

Typing

```
The next sentence will be bold. \fontsentence{\bf}This
is a bold sentence. And this one is not.
```

produces

> The next sentence will be bold. **This is a bold sentence**. And this one is not.

A space is normally suppressed after a command name, but not after delimiting characters in the macro expansion. Hence, if the ⟨argument list⟩ ends with a delimiting character, space will not be suppressed. This gives another approach to the problem of spaces after macros, discussed in §6.1

EXAMPLE 6.12 ▷ Defining

```
\def\EEF${Exotic and Edible Foods}
```

and typing

```
\EEF$ is a current leader in nutrition.
Digestron is a trademark of \EEF$.
```

produces

Exotic and Edible Foods is a current leader in nutrition. Digestron is a trademark of Exotic and Edible Foods.

In the first line of the example, the space is retained since it does not follow the ⟨macro name⟩. The dollar sign separator `$` disappears since it is not repeated in the ⟨macro text⟩ of the definition of `\EEF`.

6.4 Examples: Macros in Text Mode

This section presents a collection of additional examples.

- Separating rule

The horizontal rule command `\hrule` draws a line across the page (see §2.5). This line is not surrounded by extra space since the space is not always needed. You may define the `\ruleseparator` command as follows:

```
\def\ruleseparator{\bigbreak\hrule\medskip\nobreak}
```

This definition ensures that the rule is surrounded by a big breakable space above and by a medium unbreakable space below. The macro can be used to separate segments of text or sections of a document.

- Math symbols in text

You can borrow symbols from the math mode. If a document contains a large number of, say, heart suit symbols, it would be easier to abbreviate `$\heartsuit$` as, for instance, `\love`:

```
\def\love{$\heartsuit$}  I \love\ New York.
```

- German Hyphenation

Macros simplify entering words with unusual hyphenation. Certain German words such as "Becker" change their spelling when hyphenated (to Bek–ker, see §3.10). To ensure correct hyphenation, you should enter

```
Mr.~Be\discretionary{k-}{k}{ck}er lost the tournament.
```

instead of

```
Mr.~Becker lost the tournament.
```

Define

```
\def\ck{\discretionary{k-}{k}{ck}er}
```

in order to be able to enter the previous example as

```
Mr.~Be\ck er lost the tournament.
```

or

```
Mr.~Be{\ck}er lost the tournament.
```

- Letters and Forms

If you are typesetting a letter or a form that will be mailed to several addresses, you may want to enter the names and addresses as macros. You can then typeset different versions of a letter by simply changing the definitions of the macros. Here is a fragment of a universal recommendation letter, successfully used by the author for years:

```
\def\fellow{Mr.~John Doe}    % substitute correct name
\def\many{two}
\def\good{mediocre} % or average, very good, brilliant
\def\extra{prank} % or {great breakthrough}
\def\verymuch{with big reservations} % or {highly}
     ...
It is my pleasure to write for \fellow.
I have known \fellow\ for \many\ years. His work has
brought him a well-deserved reputation as a \good\
 researcher. His \extra s will always be remembered at
 our institution. \par
Overall I recommend him \verymuch.
```

6.5 Examples: A Simple Stylesheet

The macros discussed in this section can be combined as a simple stylesheet file. Such stylesheets can greatly simplify typesetting of books similar to the one you are reading now.

This sample stylesheet is designed for manuals that are divided into chapters and sections. The `\chapter` macro command typesets chapter titles, and `\section` typesets section titles. There are additional macros to typeset such tags as **NOTE** ▷, **TOPICS** ▷, **SYNTAX** ▷, and **EXAMPLE** ▷. Global variables are used to keep track of the numbering of the examples.

The following are the definitions of the global variables:

```
\newcount\examno % The example counter
\newtoks\chaptag % The chapter tag
```

The stylesheet will start with a few font definitions. We will define two text fonts (regular and emphasized) and two title fonts (for sections and chapters). In most cases it would be best to take these fonts from a single font family:

```
\font\text        = cmr10  % Roman at 10 points
\font\em          = cmsl10 % Slant at 10 points
\font\sectionfont = cmbx12 % Boldface at 12 points
\font\chapterfont = cmbx17 % Boldface at 17 points
```

A VTEX user may prefer to define only the `\text` font and enter the other fonts as attributes, when needed:

```
\font\text        = mvtmsr % Roman at 10 points
\define\em        = {\Slant 167}
\font\sectionfont = {\Fontscale 12 \Bold 10}
\font\chapterfont = {\Fontscale 17 \Bold 12}
```

The sections in our stylesheet are not put on separate pages, but instead are separated by a horizontal rule. Page breaks are not allowed between the rule and the title. The section titles are centered and are set in `\sectionfont`:

```
\def\section#1{
  \bigbreak\hrule\nobreak\medskip
  \centerline{\sectionfont \S #1}
 \medskip\nobreak
  }
```

The `\section` macro has a single argument. You can use `\section`, for instance, by typing

```
\section{12.3 Starting your own Grue farm.}
```

In our stylesheet, the sections are not numbered automatically.

The definition of `\chapter` is as follows:

```
\def\chapter#1.#2{
  \message{(Working on Chapter #1)}
  \vfill\eject             % start on new page
  \footcount=1             % start footnotes with 1
  \examno=1                % start examples  with 1
  \chaptag={#1}            % remember chapter number
  \leftline{\chapterfont Chapter \chaptag.\ #2}
  \medskip
  }
```

The `\chapter` command

- displays a message, notifying you that TeX has just started a new chapter (see `\message`).
- starts the chapter on a new page. It is assumed that the document does not start immediately with Chapter 1; there should be title pages. If there are no pages before Chapter 1, the first page will be skipped.
- sets footnote numbering to 1.
- sets example numbering to 1.
- records the chapter number in `\chaptag` for use in example numbering.
- left-aligns the chapter title, precedes it with the word "Chapter", and sets the title in bold.

The first argument of `\chapter` is separated by a period, the second is put in braces. You may use the macro as in

```
\chapter 12.{Grue Farms}
```

The stylesheet also contains the macro definitions for `\note`, `\topics`, and `\example`. These commands produce tags similar to the ones used in this book.

The `\note` macro that typesets **NOTE**'s is defined as follows:

```
\font\br=mvmelb scaled 1200 smallcaps
\def\note{
   \smallbreak
   \noindent{\br Note~$\triangleright$\qquad }
   }
```

The `\topics` macro that typesets **TOPICS** headers is

```
\def\topics{
  \medskip\nobreak   % no break before topics line
  \noindent{\br Topics~$\triangleright$\qquad}
  }
```

Notice the difference between these two macros: the `\note` macro is usually used in the middle of sections. A page break just before a **NOTE** is always acceptable and usually welcome. At the same time, the `\topics` macro is used immediately after the title of a section; a page break before it would be unacceptable. You can also suppress page breaks after the **TOPICS** line by changing the definition of `\topics` to

```
\def\topics#1\par{
  \medskip\nobreak   % no break before topics line
  \noindent{\br Topics~$\triangleright$\qquad#1\par}
  \smallskip\nobreak
  }
```

The \example macro is slightly more complicated, since it numbers the examples automatically. The definition of \example is

```
\def\example{
    \smallbreak % break before is OK
    \noindent
    {\br Example~\chaptag.%
       \the\examno~$\triangleright$\qquad}
    \advance\examno by 1
    }
```

Each time it is invoked, the \example macro automatically increments the example counter.

Another useful command in the manual stylesheet is the \option macro that typesets its argument in angle brackets, as is done in the syntax descriptions in this book.

```
\def\option#1{\hbox{$\langle$#1$\rangle$}}
```

Notice that the macro uses the \langle and \rangle commands rather than the character '<' and the character '>'.

6.6 Examples: Macros in Math Mode

This section describes several macros that may be used for typesetting mathematical expressions.

The following macro defines a command similar to \choose (see §4.5), but which uses brackets instead of parentheses:

```
\def\bracedchoose{ \atopwithdelims[] }
```

EXAMPLE 6.13 ▷ Typing

```
$$10121 \bracedchoose 999$$
```

results in

$$\begin{bmatrix}10121\\999\end{bmatrix}$$

You can define macros to typeset additional function names similar to the function name commands discussed in §4.6.

EXAMPLE 6.14 ▷ Typing

```
\def\trace{\mathop{\rm trace}\nolimits}
```

defines the `\trace` command that typesets the name of the "trace" function in roman. You can type

```
$$a=\trace(M)$$
```

to produce

$$a = \mathop{\rm trace}(M)$$

EXAMPLE 6.15 ▷ The `\determinant` macro operates like the `\matrix` and the `\pmatrix` commands (see §4.12). Typing

```
\def\determinant#1{\left\vert \matrix{#1} \right\vert}
$$\determinant{\cos(\phi)&\sin(\phi) \cr
   \sin(-\phi)&\cos(\phi) }$$
```

produces

$$\left\vert \begin{matrix} \cos(\phi) & \sin(\phi) \\ \sin(-\phi) & \cos(\phi) \end{matrix} \right\vert$$

You can automatically number equations with the `\neqno` macro. This macro requires a reserved count variable. Define

```
\newcount\eqncount  % A count variable for equations
\eqncount=1         % initialized to 1
\def\neqno{
  \eqno(\the\eqncount)
  \global\advance \eqncount by 1 }
```

The `\neqno` macro does not have arguments; like `\eqno`, which it uses, `\neqno` aligns the equation number on the right.

EXAMPLE 6.16 ▷ Double and triple integrals that do not have limits should be kerned together for best results. Typing

```
\def\tripleint{\int\!\!\int\!\!\int}
$\Psi(X) = \tripleint \psi(x) dx$
```

produces

$\Psi(X) = \int\!\!\int\!\!\int \psi(x) dx$

6.7 Scope of a Macro Definition

TOPICS ▷ `\outer`, `\long`, `\global`

It is possible to define macros with different properties from the ones created by `\def`. They are of interest only to very experienced users.

SYNTAX ▷ `\long` ⟨macro-definition⟩

where ⟨macro-definition⟩ refers to a macro definition via `\def`, discussed above.

The `\long` prefix command allows the macro arguments to span more than a single paragraph, which ordinary `\def` macros are not allowed to do.

➡ The macro facility is a very powerful and (*alas!*) also a very dangerous tool. It is possible to create macros that do almost anything; it is also very easy to misuse them with disastrous results. TEX imposes some restrictions on the use of the `\def` command and macros in an attempt to forestall a disaster; you may impose additional restrictions or change the defaults.

Consider a possible scenario for a disaster: define the `\boldtext` macro by typing

```
\def\boldtext#1{\bf #1}
```

Like the `\bf` command, `\boldtext` typesets the affected text in boldface; unlike `\bf`, however, it requires its argument to be explicitly surrounded by braces. What happens if one of these braces is missing?

The error is not very serious if the missing brace is the left brace. Not seeing the left brace `{`, TEX would set in bold only the character that follows `\boldtext`. Eventually, TEX would find the unmatched right brace `}`, print the warning message

```
! Too many }'s.
```

and discard the extra brace.

If the right brace is missing, the consequences could be more serious. Prior to doing anything else, TEX must locate the brace to expand the macro. A macro under expansion stays in memory; it is possible that TEX will have to read the whole of the TEX document into memory, searching for a right brace which is not there! A likely result of such a search will be the error message "`out of memory`". (Do not worry, this will not happen.)

A missing brace is a very frequent and natural error; TEX ought to be able to handle it without collapsing. The rule which does not allow a macro to span more than a single paragraph safeguards against missing braces. Because of this rule, TEX does not read beyond a single paragraph in the pursuit of a missing brace. Once the end of a paragraph is seen and the brace is not found, you get the error message

```
! Paragraph ended before \boldtext was complete.
```

If you press the carriage return, the right brace will be inserted automatically at the end of the paragraph.

If it is absolutely essential to allow the macro arguments to span more than a single paragraph, `\long\def` should be used. One should be very careful in using `\long` macros, since syntax errors in using `\long` macros may cause TeX to run out of memory. The `\long` form of the above example is

```
\long\def\boldtext#1{\bf #1}
```

SYNTAX ▷ `\outer`⟨macro-definition⟩

where ⟨macro-definition⟩ refers to a macro definition via `\def`.

Macros defined using `\outer` may not appear inside other macros or be invoked inside groups.

TeX allows the free nesting of macros. Some macros, however, should never be nested inside other macros. For example, if the `\Section` macro, defined below, is found inside another macro, it is likely to be the result of a user error. The macro designer should prefix such macros with the `\outer` prefix to simplify and speed up the search for errors.

EXAMPLE 6.17 ▷ Typing

```
\outer\def\Section#1#2.{\bigskip\rm\S#1\bf#2\medskip.}
```

defines a macro that can be used for section headers. It cannot be used inside other macros or commands. If an outer macro appears inside another command (`\underbar`, for instance), you get the error message

```
! Forbidden control sequence found while scanning use of \underbar.
```

SYNTAX ▷ `\global`⟨macro-definition⟩

where ⟨macro-definition⟩ refers to a macro definition via `\def`, as discussed above.

As mentioned above, macro definitions are local to the groups in which they are defined. The `\global` prefix enables the global declaration of macros. A global declaration stays active after the group containing it is closed.

In general, it is best to declare local macros inside groups, so they are forgotten after the group is typeset and do not waste memory.

TeX allows us to abbreviate `\global\def` into `\gdef`.

➡ The prefixes `\outer`, `\long`, and `\global` can be used together in the same definition.

6.8 Controlling Macro Expansion

TOPICS ▷ `\edef`, `\xdef`, `\noexpand`, `\expandafter`

The previously discussed forms of macros copy their arguments verbatim, without any expansion. In some applications we would prefer to have the arguments expanded immediately. Such would be the case when the macro is used to build a page reference, as in the following example:

EXAMPLE 6.18 ▷

```
 begin \def\refbegin{\the\pageno}
 is a keyword that begins statements.
............
 end (see "begin", page \refbegin)
 ends statements that begin with begin .
```

Assume that the first fragment of the text occurs on page 9 and the second on page 13. The `\refbegin` macro in the second fragment will expand into 13, not 9. This is because `\the\pageno` in the definition of `\refbegin` will be expanded only when `\refbegin` is used in the second fragment.

If expansion has to be done immediately, use the `\edef` form of the macro definition. The syntax of `\edef` is identical to the syntax of `\def`.

The `\edef` macro definition can be preceded with `\outer`, `\long`, or `\global`, as defined above. In the last case, the `\xdef` abbreviation can be used instead of `\global\edef`.

In some cases it may required to expand some portions of macro, while leaving the other unexpanded. In such cases, use the `\edef` or `\xdef` forms of the macro definition and precede the tokens that should not be expanded with the `\noexpand` command.

SYNTAX ▷ `\expandafter`

In other cases, it may be desirable to delay the expansion of certain parts of an expression. For example, typing

```
\uppercase{\romannumeral 101}
```

will not produce 'C1', since `\uppercase` will be applied before `\romannumeral` had a chance to build 'c1'. The user should define a `\upr` macro:

```
\def\upr#1{\uppercase\expandafter{#1}}
```

and use

```
\upr{\romannumeral 101}
```

instead of an above construct. Here `\expandafter` reverses the order of macro expansion.

\upr is defined in VTeX(see §11.9)

6.9 Conditional Processing: if

TOPICS ▷ `\ifodd`, `\ifnum`, `\ifdim`, `\iftrue`, `\iffalse`, `\ifeof`, `\ifvmode`
▷ `\ifhmode`, `\ifmmode`, `\ifinner`, `\ifvoid`, `\ifvbox`, `\ifhbox`
▷ `\ifx`, `\else`, `\fi`

TeX supports a variety of commands for conditional text processing. These commands are primarily designed for use in macros.

SYNTAX ▷ ⟨if operator⟩ ⟨condition⟩ ⟨true text⟩ `\else` ⟨false text⟩ `\fi`,

where ⟨if operator⟩ is one of the predefined `\if` operators: `\ifodd`, `\ifnum`, `\ifdim`, `\iftrue`, `\iffalse`, `\ifeof`, `\ifvmode`, `\ifhmode`, `\ifmmode`, `\ifinner`, `\ifvoid`, `\ifvbox`, `\ifhbox`, `\ifx`, or a new `\if` operator that was created with the `\newif` command (see §6.11). The syntax of the ⟨condition⟩ depends on ⟨if operator⟩. If ⟨condition⟩ is true, ⟨true text⟩ is processed; otherwise ⟨false text⟩ is processed. The `\else` command may be omitted if there is no ⟨false text⟩; `\fi` is always needed.

The `\ifodd` command tests if a number is odd.

SYNTAX ▷ `\ifodd` ⟨number⟩ ⟨true text⟩ `\else` ⟨false text⟩ `\fi`

where ⟨number⟩ refers to a numeric constant or a count variable.

This test is primarily used for distinguishing odd-numbered and even-numbered pages in the document.

EXAMPLE 6.19 ▷ Typing

```
This page (\the\pageno) is
 \ifodd\pageno{odd}\else{even}\fi-numbered.
```

produces

This page (171) is odd-numbered.

One can incorporate the results of the test into the formatting of `\footline` or `\headline` to format odd-numbered and even-numbered pages differently.

EXAMPLE 6.20 ▷ Typing

```
\footline={\ifodd\pageno{\hss\rm\folio}
   \else{\rm\folio\hss}\fi}
```

forces page numbers on even pages to be left-justified and page numbers on odd pages to be right-justified. (See also `\footline`, `\headline`, `\pageno`, and `\folio`.)

The `\ifnum` command compares two numbers.

SYNTAX ▷ `\ifnum` ⟨number$_1$⟩ ⟨comparison⟩ ⟨number$_2$⟩ ⟨true text⟩ `\else` ⟨false text⟩ `\fi`

where ⟨comparison⟩ is one of '=', '<', and '>'; ⟨number$_1$⟩ and ⟨number$_2$⟩ refer to numbers or count variables.

EXAMPLE 6.21 ▷ This example shows how to format a newsletter headline making special provisions for the New Year Day. Typing

```
\centerline{\EEFlogo\ News:
\ifnum\day=1{\ifnum\month=1{New Year Edition}
 \else { Daily Edition: \the\month/\the\day}\fi}
\else{ Daily Edition: \the\month/\the\day}\fi}
```

on the day when this book was last typeset produced

$\mathcal{EEF}$ News: Daily Edition: 8/2

The `\ifdim` command is similar to `\ifnum`; it compares dimensions instead of numbers.

SYNTAX ▷ `\ifdim` ⟨dimen$_1$⟩ ⟨comparison⟩ ⟨dimen$_2$⟩ ⟨true text⟩ `\else` ⟨false text⟩ `\fi`

EXAMPLE 6.22 ▷ The following simple program interacts with the user, reminding him of the necessity to conserve toner and paper:

```
\ifdim\baselineskip>1in
     {\errmessage{Hey, you are wasting valuable supplies!}
\else{\message{Good, you listened!} \fi
```

The `\message` and `\errmessage` commands are explained in §7.1.

The operators `\iftrue` and `\iffalse` do not perform actual tests; they return true and false, respectively.

SYNTAX ▷ `\iftrue` ⟨true text⟩ `\else` ⟨false text⟩ `\fi`

The `\ifeof` command tests if an input file is being read. It is explained fully in §7.2.

SYNTAX ▷ `\ifeof` ⟨number⟩ ⟨true text⟩ `\else` ⟨false text⟩ `\fi`

where ⟨number⟩ refers to an `input` file (0 through 15). `\ifeof` returns false only if the file is open and has not been completely read.

The `\ifhmode`, `\ifvmode`, `\ifmmode`, and `\ifinner` ⟨if operators⟩ can be used to fine-tune macros whose behavior depends on the mode of operation (see §8.1 for more details). The `\ifvmode` command tests if the program is in a vertical mode; the `\ifhmode` command tests if the program is in a `horizontal mode`. The `\ifmmode` command tests if the program is in a `math mode`. The `\ifinner` command tests if the program is in an `internal mode` (restricted horizontal, internal vertical, or ordinary math, see §8.1).

SYNTAX ▷ `\ifvmode` ⟨true text⟩ `\else` ⟨false text⟩ `\fi`
▷ `\ifhmode` ⟨true text⟩ `\else` ⟨false text⟩ `\fi`
▷ `\ifmmode` ⟨true text⟩ `\else` ⟨false text⟩ `\fi`
▷ `\ifinner` ⟨true text⟩ `\else` ⟨false text⟩ `\fi`

EXAMPLE 6.23 ▷ The `\determinant` macro, described in §6.6, should not be used in non-math modes. The following definition forces the display math mode if `\determinant` is invoked in a non-math mode.

```
\def\determinant#1{
  \ifmmode       \left\vert \matrix{#1} \right\vert
    \else     $$ \left\vert \matrix{#1} \right\vert $$
    \fi }
```

Instead of fixing the user's error by going into the display math mode, you may simply inform him about the error with the `\errmessage` command (see §7.1).

There are other ⟨if operators⟩ described later in the book. The `\if` and `\ifcat` ⟨if operators⟩ are in §6.18. The `\ifvoid`, `\ifhbox`, `\ifvbox`, and `\ifx` are described in §8.4.

6.10 Conditional Processing: case

TOPICS ▷ `\ifcase`, `\or`, `\else`, `\fi`

The `\ifcase` command is similar to Pascal's `case` or la C's `switch`. It selects a required possibility out of many choices without making repeated `\if` tests.

SYNTAX ▷ `\ifcase` ⟨number⟩ ⟨text for number=0⟩ `\or` ⟨text for number=1⟩ `\or` ...`\or` ⟨text for number=N⟩ `\else` ⟨default text⟩ `\fi`

Like the `\if` commands, `\ifcase` allows one to omit `\else` if there is no ⟨default text⟩.

EXAMPLE 6.24 ▷ The following macros print the current month and the date symbolically:

```
\def\monthname{\ifcase\month
 \or{January}  \or{February}  \or{March}
 \or{April}    \or{May}       \or{June}
 \or{July}     \or{August}    \or{September}
 \or{October}  \or{November}  \or{December}\fi}
\def\today{\monthname \the\day}
```

Typing the above definitions and

```
Today is \today.
```

on the day this book was last typeset, produced

Today is August 2.

`\today` is standard in VTEX. Also defined is `\dayofweek`, see §11.9.

EXAMPLE 6.25 ▷ Assume that you have designed a `\random` macro which randomly returns numbers between 0 and one less than its argument. You now can create a simple generator of random responses:

```
\def\excuse{\ifcase\random 5
 {a large number of orders} \or
 {the secretary's error} \or
 {our desire to bring you the best possible product} \or
 {a computer failure} \or
 {terrorists' activity in the Middle East} \fi
```

and use it as follows:

```
Dear Mr.~Smith. We are happy to hear from you yet again.
Unfortunately, we could not fill your order because of
\excuse. Please stay in touch.
```

`\random` is standard in VTEX.

6.11 Conditional Processing: User–Defined Tests

TOPICS ▷ `\newif`

The `\newif` command creates new forms of the `\if` statement.

SYNTAX ▷ `\newif \if` ⟨new if test name⟩,

where ⟨new if test name⟩ is the name of the new `\if` operator. `\newif` will automatically append `\if` to the beginning of ⟨new if test name⟩ and also create two more commands by appending `true` and `false` to the end of ⟨new if test name⟩.

EXAMPLE 6.26 ▷ Typing

```
\newif\my
```

creates the commands `\ifmy`, `\mytrue`, and `\myfalse`.

User-defined if commands are analogous to boolean or logical variables in programming languages. When created, the ⟨if test⟩ variable has the value "false"; to set it to "true", use the `\mytrue` command; use `\myfalse` to set it to "false".

EXAMPLE 6.27 ▷ Input

```
\newif\my
\mytrue \ifmy\message{(TRUE)}\else\message{(FALSE)}
\myfalse\ifmy\message{(TRUE)}\else\message{(FALSE)}
```

would print `(TRUE)` and then `(FALSE)` to the screen.

6.12 Conditional Processing: loops

TOPICS ▷ `\loop`, `repeat`

TeX's `\loop` and `\repeat` macros define a primitive mechanism which emulate `repeat...until` and `for` loops.

SYNTAX ▷ `\loop` ⟨text⟩ ⟨if test⟩ `\repeat`

where ⟨text⟩ is any sequence of commands and characters; ⟨if test⟩ is one of the `\if` commands defined above.

EXAMPLE 6.28 ▷ The following commands will cause numbers 1 through 100 to be echoed on the screen:

```
\newcount\cnt \cnt=1
\loop
  \message{\the\cnt}
  \advance\cnt by 1
  \ifnum\cnt<101
  \repeat
```

Notice that `\loop...\repeat` executes at least once.

6.13 Category Codes

TOPICS ▷ character categories, category codes

When TEX processes an input file, it treats each character according to its category code. There are 16 category codes (0 through 15). Some of the codes correspond to characters that will actually appear in the manuscript (category code 11 corresponds to letters); some are part of the syntax (category 3 contains the math shift character $).

TEX predefines category codes for all characters that can be produced on the keyboard; all information in this book assumes that these codes have not been changed. However, sometimes it is convenient to re-assign the category codes.

The 16 categories are defined as follows:

Cat Code	Category Name
0	Escape Operator
1	Beginning of Group Operator
2	End of Group Operator
3	Math Shift Operator
4	Alignment Tab Operator
5	End of Line Character
6	Macro Parameter Operator
7	Superscript Operator
8	Subscript Operator
9	Ignored Character
10	Space Character
11	Letter
12	Other Character
13	Active Character
14	Comment Operator
15	Invalid Character

The meaning of each category is as follows:

0 *Escape:* The escape operator begins a command, which is composed of upper- and lower-case letters. By default, the '\' acts as the escape. The escape operator and the command that follows it are not typeset but interpreted by TEX. The character '\' can be produced by the `\backslash` command (in math mode).

1,2 *Beginning of Group and End of Group:* By default, the left brace '{' opens groups in TEX and the right brace '}' closes them. To typeset '{' and '}', use the `\{` and `\}` commands in math mode.

3 *Math Shift:* The math shift operator, '$' by default, opens and closes the ordinary math and the display math modes. Portions of text that should appear in math mode should be surrounded by the math shift operator; portions of text that should appear as display math should be surrounded by the math shift operator written twice. To typeset '$', use the `\$` command in any mode.

4 *Alignment Tab:* The alignment tab operator is used in tables and matrices to control the alignment of the text. By default it is '&'; to typeset '&', use the `\&` command.

5 *End of Line:* The end of line character (default ASCII 13 or Ctl-M) is an invisible character which is stored in ASCII files to separate lines of text. For MS-DOS, it is actually a pair of characters (CR and LF – carriage return and line feed) but TEX always views end of line as a single character EOL, and translates it into a space unless `\obeylines` is in effect.

6 *Macro Parameter:* The macro parameter operator is used to specify the arguments of a macro. It defaults to '#'; to typeset '#', use the `\#` command in any mode.

7,8 *Superscript and Subscript:* These operators,^ and _ by default, are used in math mode to create superscripts and subscripts. To typeset _, use the `\_` command in math mode.

9 *Ignored:* Ignored characters are not processed by TEX. By default, no characters are ignored.

10 *Space:* Space is never typeset explicitly, since the device drivers use the printer's horizontal motion commands for precise positioning. Space characters are used to separate text in the source file but are always ignored after commands. For instance, "`\TeX␣␣␣,`" is identical to "`\TeX,`". Leading spaces at the beginning of the line are always ignored. Spaces become significant if `\obeyspaces` is in effect. The printable space character '␣' can be produced, for example, as `{\tt\char32}`. By default, the tab character (ASCII 9) is treated as a space.

11,12 *Letter and Other:* By default, letters are lowercase and uppercase letters. "Other" characters include all the characters not placed in other categories, i.e., digits 0 through 9 and most of the punctuation signs. Both letters and other characters are processed and typeset by TeX. However, letters may appear in command names while other characters may not. Thus, if TeX sees `\hskip1pt`, it processes the command `\hskip`, followed by the characters `1`, `p`, and `t`.

13 *Active Character:* An active character is a character that is viewed by TeX as a command. By default, `~` is an active character, with the meaning: a thin unbreakable space. Active characters can be used in macro definitions. For instance, you can define `\def~{\message{Hello}}`.

14 *Comment:* The comment character, % by default, specifies the beginning of a comment. To typeset '%', use the `\%` command.

15 *Invalid:* Invalid characters are not processed or typeset by TeX, but they cause an error message to be printed. Most ASCII characters that cannot be entered directly from the keyboard are viewed as invalid. For example, if the source file contains an ASCII 127, you will receive an error message.

The category codes can be examined with the `\show` command and changed by `\catcode`. However, do not change them frivolously. Some examples of possible changes are given with the description of the `\catcode` control sequence. This book assumes that the category codes have not been altered; if they have been, the behavior of the package will change radically.

EXAMPLE 6.29 ▷ If the character **$** were assigned to category 11 (Letter), it would be possible to use it as part of the name of a control sequence; however, it would no longer be possible to enter math mode. If now one were to assign the category 4 code (math shift) to some other character, say `/`, this character would act as a math shift, delimiting mathematical expressions. If this were indeed done (*don't!*), all the references in this book to the math shift character **$** should be treated as references to the new math shift character `/`.

TeX usually translates the `EOL` (end-of-line) character into a space. In some cases this might be inconvenient. If it is desirable to preserve the original line breaks, the command sequence `\obeylines` should be issued. If it is necessary to completely ignore the `EOL` character and not insert a space, the source line should end with the comment character (`%`).

EXAMPLE 6.30 ▷ Typing

```
Mr.~Typo—one of the higher paid typists in \EEFlogo—
was very happy to discover \TeX
```

results in

Mr. Typo—one of the higher paid typists in $\mathcal{EEF}$— was very happy to discover TEX

To suppress the space before "was", enter the first line of the example as

```
Mr.~Typo-one of the higher paid typists in \EEFlogo-%
```

6.14 Changing Category Codes

TOPICS ▷ `\catcode`

You can change category codes to suit unusual applications. This feature should be exercised with restraint, since modifications on this level may cause unpredictable results.

SYNTAX ▷ `\catcode` ⟨character⟩ [=] ⟨category code⟩

where ⟨category code⟩ is as defined in the previous section and ⟨character⟩ is a number between 0 and 127 (only lower ASCII codes are allowed).

EXAMPLE 6.31 ▷ Typing `\catcode`\[=1` (see §5.2 for an explanation of the number format) makes the left bracket ( [) act as-a-beginning of group operator. If you also type `\catcode`\]=2`, you can use brackets instead of braces to specify groups. To typeset the brackets, you will have to use the `\lbrack` and `\rbrack` commands.

EXAMPLE 6.32 ▷ The following sequence of instructions temporarily redefines $ to be a letter to create a private control sequence. Typing

```
\catcode`\$=11 \def\${Hello} "\$ World" \catcode`\$=3
```

produces

"Hello World"

It is possible to redefine the category codes of all reserved characters (#, $, etc) and use the ASCII control codes (Ctl-A, Ctl-B, etc.) instead to denote special-purpose operators.

TEX defines the control sequence `\active` to mean "13". No control sequences are defined for other category codes. As will be seen from the next section, the most commonly used change of the category is to `\active`.

6.15 Using High ASCII Characters

Version 3.0 of TEX language, introduced in 1990, allows use of ASCII characters 128-255. This can somewhat simplify the input of such special characters as accented letters, Greek letters, and/or line-draw characters. Unfortunately, since different type of computers use different character sets for ASCII characters 128-255, the examples in this section can only be used on an IBM PC.

Greek letters α and β appear in th IBM PC character set as characters 224 and 225. This enables us to define

```
\catcode224=\active \defα{\alpha}
\catcode225=\active \defβ{\beta}
```

and therefore type "α" and "β" instead of lengthier `\alpha` and `\beta`.

The IBM PC character set also contains a number of accented characters. For instance, character 160 is "á", normally entered as `\'a`. European users of TEX would prefer to enter "á" as a single character, especially since this would probably be required in order to use spelling checkers. For this, use

```
\catcode160=\active \def\'a{\'a}
```

On a large scale, the same approach can be used with Cyrillic fonts (see §2.13).

Finally, the IBM PC character set includes a number of line-draw characters. These characters may be essential in typesetting screen dumps and certain kinds of tables. While standard TEX fonts lack line-draw characters, VTEX includes the `mvttld` font that extends the `\tt` font with line-drawing capabilities. To use this or similar font, one must first build a translation table similar to the ones given above.

6.16 Upper and Lower Cases. Conversion

TOPICS ▷ `\uppercase`, `\lowercase`, `\uccode`, `\lccode`

The `\uppercase` command changes the case of its argument's tokens to upper case; `lowercase` changes the case to lower case. The precise rules of conversions are controlled by `\uccode` and `\lccode`.

SYNTAX ▷ `\uppercas`ר ⟨argument⟩

▷ `\lowercase` ⟨argument⟩

EXAMPLE 6.33 ▷ Typing

```
Hello, \uppercase{world!}
```

results in

Hello, WORLD!

The name of the current section of a document could be capitalized with `\upper` before being used in a running head created with `\headline`. Similarly, the entries for an index can be converted to lower case with `\lowercase`.

EXAMPLE 6.34 ▷ The `\uppercase` and `\lowercase` commands can be used in conjunction with fonts that do not distinguish between uppercase and lowercase characters. Assume that the OCR10 font contains only uppercase characters. You can design a special macro to access the characters in this font:

```
\font\ocr=ocr10  \def\OCR#1{\uppercase{\ocr #1}}
```

If by mistake one tries to typeset a lowercase letter in OCR, it will be automatically converted to upper case.

TeX remembers the uppercase and lowercase case counterparts for each character. The conversion takes place if the corresponding uppercase (or lowercase) character is different from the zero character. The exact rules of conversion are given and can be changed by operators `\uccode` and `\lccode`.

SYNTAX ▷ [`\uccode` | `\lccode`] ⟨character_1⟩ [=] ⟨character_2⟩

where ⟨character_1⟩ and ⟨character_2⟩ are integer numbers between 0 and 127 (see §5.2).

By default, TeX assigns

```
\uccode'\a='\A \uccode'\b='\B  ... \uccode'\z='\Z
\lccode'\A='\a \lccode'\B='\b  ... \lccode'\Z='\z
```

and sets all other codes to 0, so there is no conversion for other characters.

EXAMPLE 6.35 ▷ If you type

```
\uccode'\A='\a  \uccode'\B='\b  ... \uccode'\Z='\z
```

the `\uppercase` command will actually "flip the case" since lowercase letters are specified as upper cases of uppercase letters. It is also possible to use the `\uppercase` command to implement a simple permutation cipher.

6.17 Accessing Individual Characters

TOPICS ▷ `\char`, `\chardef`

Fonts normally contain at most 256 characters, numbered 0 through 255. Since not all characters can be accessed from the keyboard, you may refer to characters by their position numbers.

SYNTAX ▷ `\char` ⟨number⟩, where ⟨number⟩ is between 0 and 255 (see §5.2)

The character codes are based on ASCII, but there are some differences. Most of the visible ASCII characters, i.e., those with decimal codes 33 through 126, appear in most of the fonts in the same positions. Therefore, either `\char"41` or `A` defines "A" in most of the fonts. However, the characters 0 through 32 and character 127 are different in TEX fonts. For example, in most of the standard TEX fonts, `\char1` denotes uppercase Greek Γ and `\char127` denotes the umlaut accent ¨.

Many of the TEX commands are just abbreviations that request a particular `\char`, possibly from a particular `font`. For example, `\%` refers to `\char37` in the current font. Most of these abbreviations were created using the `\chardef` command.

SYNTAX ▷ `\chardef` ⟨char name⟩ [=] ⟨number⟩

where ⟨char name⟩ is a new command that denotes `\char`⟨number⟩.

EXAMPLE 6.36 ▷ Typing

```
\chardef\umlaut=127
```

defines a new command `\umlaut` that accesses `\char127` in the current font. Then typing

```
A\umlaut B\umlaut C\umlaut D
```

results in

A¨B¨C¨D

6.18 Tokens. Token Strings

TOPICS ▷ `\show`, `\if`, `\ifcat`, `\string`, `\csname`, `\endcsname`, `\newtoks`

TEX views the input as a sequence of tokens. A token is either a character in one of the sixteen character categories or a command. The `\show` command displays the information about a token.

SYNTAX ▷ `\show` ⟨token⟩

EXAMPLE 6.37 ▷ In interactive mode, typing

```
*\show A
```

causes TEX to print

```
> the letter A.
```

Similarly, typing

```
*\show \bigskip
```

returns

```
> \bigskip=macro:
->\vskip \bigskipamount .
```

The `\if` ⟨if operator⟩ compares two tokens that follow it.

SYNTAX ▷ `\if` ⟨token$_1$⟩ ⟨token$_2$⟩ ⟨true text⟩ `\else` ⟨false text⟩ `\fi`

The `\ifcat` ⟨if operator⟩ compares the category codes of two tokens that follow it.

SYNTAX ▷ `\ifcat` ⟨token$_1$⟩ ⟨token$_2$⟩ ⟨true text⟩ `\else` ⟨false text⟩ `\fi`

⟨true text⟩ will be typeset if ⟨token$_1$⟩ and ⟨token$_2$⟩ are of the same category.

A command name is viewed as a single token. The `\string` command makes `\TeX` interpret it as a sequence of the individual characters that form the name.

SYNTAX ▷ `\string` ⟨command⟩

EXAMPLE 6.38 ▷ Typing

```
\tt \string\bigskip \string\TeX
```

produces

```
\bigskip\TeX
```

The `\tt` font was used since the backslash character does not appear in most standard TEX fonts. In the `\rm` font, for instance, it would be typeset as “. VTEX handles the backslash character correctly.

Conversely, the \csname and \endcsname commands form a command from individual characters (tokens).

SYNTAX ▷ \csname ⟨tokens⟩ \endcsname

All the tokens between \csname...\endcsname are compressed together to make a command sequence. \csname...\endcsname are essential when one needs to dynamically build names of control sequences. An example of this is a cross-reference facility (see §7.3).

The ⟨toks⟩ variables are used to hold strings of tokens. ⟨toks⟩ variables are similar to parameterless macros.

Two important ⟨toks⟩ variables are \headline and \footline (see §3.3).

Like other types of variables, ⟨toks⟩ variables can be accessed with commands \the and \showthe. ⟨toks⟩ variables can be set by an assignment. New ⟨toks⟩ variables can be created with the \newtoks command.

256 ⟨toks⟩ variables can also be directly accessed with the \toks command. Changes to ⟨toks⟩ variables made inside a group are local.

6.19 Predefined Token Variables

TOPICS ▷ \output, \everypar, \everymath, \everydisplay, \everyhbox
▷ \everyvbox, \everyjob, \everycr, \errhelp

TeX predefines a number of token string variables. They are

- \output The user output routine. Writing \output routines is perhaps the most difficult part of TeX. The default \output routine appends \headline and \footline to the body of the page. Custom \output routines can be, in particular, used to post-process the page after it has been formatted. A custom \output routine would be required to implement double-column formats.
- \everypar The list of tokens to be inserted at the beginning of each paragraph.
- \everymath The list of tokens to be inserted whenever math mode is entered.
- \everydisplay The list of tokens to be inserted whenever display math mode is entered.
- \everyhbox The list of tokens to be inserted whenever an \hbox begins.
- \everyvbox The list of tokens to be inserted whenever a \vbox begins.
- \everyjob The list of tokens to be inserted at the beginning of each job.

- `\everycr` The list of tokens to be inserted after every `\cr`.
- `\errhelp` The help that supplements `\errmessage` (see §7.1)

Most of these commands are used infrequently. `\everymath` and `\everydisplay` are used in VTeX `\Fontscale` command (see §11.9). `\everypar` would be essential for numbering the paragraphs. `\everyjob` is a common device for inserting a brief copyright message:

```
\everyjob{\message{
 * * * Super Duper Macro Pack * * *
 Copyright (c) 85,86,87,88,89,90,91,92,93,94,95 and 96
 by Tex Smith. Not responsible for erased files.} }
```

6.20 Other Macro Commands

TOPICS ▷ `\let`, `\afterassignment`, `\aftergroup`

A few additional macro commands described in this section are intended primarily for the macro designer. They are rarely needed in everyday work. The first command of this group is `\let`:

SYNTAX ▷ `\let`⟨macro name⟩`=`⟨token⟩

`\let` makes the ⟨macro name⟩ behave exactly as the token it is assigned. For example, typing

```
\let\<={ \let\>=}
```

now allows you to enter

```
\centerline\<text\>
```

The `\afterassignment` and `\aftergroup` commands delay the expansion of the following token until either an assignment has been performed or a group has ended. In the case of `\afterassignment`, the token is preserved in a special location; after assignment has been completed, it is re-processed by TeX. Thus, nested `\afterassignment`'s are not possible. In the case of `\aftergroup`, the token is saved on a stack, allowing nested `\aftergroup`'s.

Chapter 7

Input / Output & Extensions

7.1 Printing Messages on Screen

TOPICS ▷ \message, \errmessage, \errhelp

The \message and \errmessage commands write output to the screen. These commands do not append a carriage return to the output as \write does (see §7.2).

EXAMPLE 7.1 ▷ Typing

```
\message{Hello, World!}
```

prints "Hello, World!" to the screen (and to the log file).

In the manual.sty sample stylesheet discussed above, the \message command is used in the \chapter macro to announce each chapter as TEX starts processing it. Since the page numbers are relayed to the screen automatically, you will know immediately on which page each chapter starts.

The \errmessage command is similar to \message but it also generates an error. It may be used inside macros to verify the correctness of the arguments. \errhelp specifies the help message that will be printed if the user requests additional help.

EXAMPLE 7.2 ▷ The following sequence of commands will prompt the user for a number, analyze the input (see \read, §7.2), and print an error message if the input is invalid. Please refer to the appropriate sections of the manual for information on the other commands used in this example.

```
\newcount \sockcount
\message{How many socks do you want to buy?\ }
\read-1 to \temp  % Cannot read directly to \sockcount
  \sockcount=\temp
\ifodd\sockcount
 \errhelp{We sell socks only in pairs.
    This time you get none.}
 \errmessage{Bad sock count}
 \sockcount=0
 \else
   \divide\sockcount by 2
   \message{You are going to get \the\sockcount\ pairs}
   \makeinvoice % Some macro that typesets an invoice
 \fi
```

7.2 File and Terminal Operations

TOPICS ▷ `\openin`, `\openout`, `\newread`, `\newwrite`, `\read`
▷ `\write`, `\immediate`, `\close`

TeX supports operations on text files. Text files may be opened, closed, read, and written from TeX. Up to 16 files can be opened for reading and up to 16 files for writing, 32 files altogether. The files are numbered 0 through 15 in both cases; read files are independent from write files, so that the input file #4 has nothing to do with the output file #4.

The `\openin` command opens a file for input; `\openout` opens a file for output.

SYNTAX ▷ `\openin`⟨file number⟩=⟨file name⟩
▷ `\openout`⟨file number⟩=⟨file name⟩

where ⟨file number⟩ is from 0 to 15 and ⟨file name⟩ is a legal file name in your operating system. If ⟨file name⟩ lacks a file extension, `.tex` will be assumed.

EXAMPLE 7.3 ▷ Typing

```
\openout 7 = myfile.doc
```

opens the file '`MYFILE.DOC`' for writing.

The `\newread` and `\newwrite` commands create file variables to avoid addressing files by numbers. For instance, typing `\newread\myfile` creates a file variable `\myfile`. `\myfile` will be assigned some valid file number. To actually open the file, use `\openin\myfile=myfile.doc`.

The `\write` command writes a line of text into an output file and the `\read` command reads a line of text from an input file. One entire line of text must be processed at a time.

SYNTAX ▷ `\write`⟨file number⟩ `{`⟨text⟩`}`

writes ⟨text⟩ into file ⟨file number⟩. A file variable may also be used instead of ⟨file number⟩. If ⟨file number⟩ was not opened or is not a valid file number (negative or greater than 15), the output is written to the log file; in this case the output is also written to the terminal unless the ⟨file number⟩ is negative.

EXAMPLE 7.4 ▷ Typing `\write 7 {Hello, World}` writes "Hello, World" to the output file #7.

EXAMPLE 7.5 ▷ Typing `\write 100 {Today is \today}` writes today's date to the log file and to the terminal; if 100 is replaced by -1, this output will go only to the log file.

SYNTAX ▷ `\read` ⟨file number⟩ `to` ⟨command⟩

`\read` creates a new command (macro) and initializes it to the contents of the current line of ⟨file number⟩. A file variable may be used instead of ⟨file number⟩. If ⟨file number⟩ is not a valid file number, the input will be read from the keyboard.

Unless the line is read from the terminal, it must be enclosed in braces that must be matched. If it contains extra left braces, additional lines of text will be read until braces match; if there are extra right braces, an error will be generated.

EXAMPLE 7.6 ▷ `\read\myfile to \newline` reads one line of text from `\myfile` (it should already be open for input) to `\newline`.

EXAMPLE 7.7 ▷ `\read-1 to \newline` reads a line of text from the terminal to `\newline`.

If `\read` is performed from the terminal, you are prompted with the name of ⟨command⟩ followed by an equal sign if ⟨file number⟩ is positive; there is no prompt if ⟨file number⟩ is negative.

The `\close` command closes input and output files. Upon the completion of a job all files are closed automatically, but it is a good idea to always close files after the I/O operations are completed.

Normally, when TeX encounters a file operation, its execution is delayed until the page that contains the operation is completely typeset. This is because execution often depends on the particulars of the page; if you type `\write\myfile{\folio}`, the value of `\folio` will not be known until page breaks are computed. If a file operation must be performed immediately, use the `\immediate` prefix with any of the file commands.

EXAMPLE 7.8 ▷ `\immediate\openin 7`

The `\ifeof` command is used in reading file to check for the end of the file. `\ifeof` always returns true for empty and non-existent files so that it may be used to verify the existence of a file.

EXAMPLE 7.9 ▷ Here is a sample TeX program that performs file operations:

```
\newwrite\test              % declare file variable 'test'
\openout\test=test.txt      % open file for writing
\write\test {{ This is a test file }}
  % Second braces in a \write will be written to a file
\write\test {{ This is page \folio }}
\write\test {{ End of test }}
```

```
\closeout\test
```

The `test.txt` file will contain three lines. The correct page number will appear in the second line.

```
{ This is a test file }
{ This is page 1}
{ End of test }
```

The braces are written to the output file to allow it to be read back into TEX. The following program reads it:

```
\newread\test
\openin\test=test.txt
\ifeof\test    % what if file is not found?
  \errmessage{file "test.txt" not found}
\else
\read\test to \a    \write100\a
     % \write100 writes to screen (and log)
\read\test to \a    \write100\a
\read\test to \a    \write100\a
\closein\test
\fi
```

During file reading, use the `\ifeof` command to check the existence of the read file and the end-of-file condition.

7.3 Text References

One of the most important applications of the TEX file mechanism is the ability to resolve text references. Assume that the source includes the following two fragments of text:

```
% Fragment 1
Gnue farming (see {\it Gnues}, page ?)
is a proper activity for some gentlemen.
.........
% Fragment 2
Gnues are nocturnal animals with large fins.
```

We would like to replace the question mark in the first fragment with the actual number of the page containing the second fragment. One solution is to typeset the document, look up the required page numbers, insert them back into the source and typeset the document again; this solution requires lots of manual work and readjustment every time page numbers change. Another solution is available if the first fragment actually follows the second: we can append

```
\edef\gnueref{\the\pageno}
```

to the second fragment of text; in the first fragment we will change the question mark to `\gnueref`. This will work as long as the document does not contain any forward references.

The universal solution requires two-pass processing of the document. On the first pass we shall write the references to a special reference file. On the second pass we will `\input` the reference file. The following macros are defined:

```
\def\startreferences{
\input \jobname.ref
\newwrite\out
\immediate\openout\out=\jobname.ref}

\def\makeref#1
  {\immediate\write\out
    {\def\csname ref#1\endcsname{\number\pageno}}
  }

\def\useref#1
   {\csname ref#1\endcsname}
```

Here `\startreferences` is used once at the beginning of the document. It re-reads a previously prepared list of references; it also opens a new reference file. This file will have the same name as the document (`\jobname`) and extension `.ref`; TEX will know it as the `\out` file. The `\makeref` macro is used to place references in the reference file; `\useref` is used to retrieve the reference. Taking these macros into account, the example above should be rewritten as

```
% Fragment 1
Gnue farming (see {\it Gnues}, page \useref{Gnue})
is a proper activity for some gentlemen.
.........
% Fragment 2
Gnues are nocturnal animals with large fins.
\makeref{Gnue}
```

7.4 Indices

Output files can also be utilized to assist in preparation of indices. TEX, however, cannot build a full index by itself, since preparation of an index requires sorting the index data. Typically, the job of index preparation can be broken into three stages:

- TEX writes index data into an index files, using a macro similar to `\makeref` in the previous section.

- Index is sorted and re-arranged, either by hand or with an auxiliary program; the results are written into an include file.
- TeX formats the include file obtained in the previous section.

In most cases, it is the second step that is the most time consuming. Some authors prefer to do the entire formatting themselves; others prefer to use an auxiliary sort utility, such as the VTeX's IDXSRT program. In most cases generation of an index will require running TeX twice, with an intermission devoted to sorting. A major time-saver in the case of VTeX is the `\exec` primitive, described in the next section.

For a full set of index macros, see §11.9.

7.5 Running Other Tools from within TeX.

TOPICS ▷ `\exec`, `\command`, `\errno`

✔ This section is applicable to VTeX only.

The `\exec` and the `\command` commands allow execution of other programs from within TeX. `\exec` and `\command` can supplement TeX with abilities that are absent in TeX but present in other programs.

SYNTAX ▷ `\exec{`⟨name of program⟩`}{`⟨argument string⟩`}`
▷ `\command{`⟨command string⟩`}`

`\command` expects a single argument, entered as if it were printed at the operating system prompt. `\exec` separates the name of the command from its argument string. Under MS-DOS, there are a few subtle differences between these two forms:

- `\command` invokes MS-DOS command processor and, therefore, consumes more memory.
- `\exec` requires the full path name as the ⟨name of program⟩.
- `\command` can execute internal commands, such as `dir` or `copy`.

EXAMPLE 7.10 ▷ The interface with an indexing program (see the previous section) can be arranged as follows:

```
 \newwrite\out
 \immediate\openout\out=temp.idx
...
( write the index )
...
 \immediate\closeout\out  % close index file
 \exec{idxsrt.exe}{temp.idx \jobname.idx}
```

```
\command{erase temp.idx}
\centerline{Index}
\input \jobname.idx
\command{erase \jobname.idx}
```

Here TEX prepares a raw index in the `temp.idx`. Once the index is completed, TEX closes the file and invokes the IDXSRT program that sorts the index and writes it back into `\jobname.idx`. TEX deletes the temporary file, reads (and typesets) the sorted index, and finally deletes it.

The `\errno` integer variable is used to report back the results of `\exec` or `\command`. The value of 0 indicates a success; other values are error codes set by the operating system.

7.6 Special Extensions

TOPICS ▷ `\special`

The `\special` command allows you to copy a string verbatim to the `.dvi` file.

SYNTAX ▷ `\special` ⟨argument⟩,

where ⟨argument⟩ consists of the text that should be copied to the `.dvi` file.

EXAMPLE 7.11 ▷ `\special{"Hello"}` copies the string `"Hello"`, including quotes, verbatim to the `.dvi` file.

The interpretation of the text is done by the device drivers. There is no current standard for the syntax to be used in `\special` strings; one generally hopes that the drivers will recognize the special commands that relate to them and ignore the ones that they cannot understand.

Some examples of `\special`'s follow:

- Many drivers use `\special{ps:...}` and `\special{hp:...}` commands to indicate commands for either PostScript or PCL (HP LaserJet) printers.
- Extensions of `\TeX` designed for color printing use `\special` to transmit color-change commands.
- VTEX uses special commands that immediately follow font-description commands to transmit additional information about fonts, such as slant, aspect, and other font effects.
- VTEX also uses special commands to place bookmarks in the document; the previewer can rapidly find a bookmark.

Most device drivers come with the manufacturer's information about the list of the supported `\special` commands.

Chapter

8

Modes Rules Boxes

8.1 Modes of Operation

TOPICS ▷ vertical mode, horizontal mode, math mode, restricted horizontal mode
▷ internal vertical mode, ordinary math mode, display math mode
▷ `\ifhmode`, `\ifvmode`, `\ifmmode`, `\ifinner`

During formatting, TeX performs a large number of different tasks. It must determine the optimal spacing between words, find the best line breaks, hyphenate some words, determine inter-paragraph spacing, and break text between pages. In addition, if mathematical formulas appear within the text, TeX must typeset them using an entirely different set of rules than when typesetting ordinary text.

The variety of tasks is divided into six modes of operation. The behavior of TeX is dependent upon its current mode; some commands make sense in one mode but do not belong in another mode.

Usually, TeX switches modes on its own and this is transparent to you. However, sometimes it is important that you understand exactly what TeX is doing.

- Horizontal modes of operation are used when TeX typesets lines of text. The less general *restricted horizontal mode* is used for typesetting single lines of text and horizontal boxes (the `\hbox` command, §8.3); *horizontal mode* is used for typesetting whole paragraphs. The primary difference is that in the horizontal mode TeX has to perform line-breaking operations that do not take place in the restricted horizontal mode.
- There are also two vertical modes of operation. The *vertical mode* is used when TeX handles problems of putting the paragraphs together and determining the page breaks. The less general *internal vertical mode* is used to do vertical operations within a paragraph (such as in a `\vbox`, see §8.3).
- Finally, there are two math modes of operation: *ordinary math mode* and *display math mode*. The first is used to typeset expressions appearing in the body of paragraphs of text; the second typesets equations that are displayed on separate lines.

Overall, the differences between similar modes (the horizontal mode and the restricted horizontal mode, the vertical mode and the internal vertical mode, the math mode and the display math mode) are slight; the differences between unrelated modes are substantial. For example, there are many commands that may be used exclusively in math mode; there are just a few that may be used in one of the math modes but not in the other (`\eqno`, `\leqno`).

A macro designer should be aware of the existence of different modes of operation. Macros can determine the mode in which they are invoked by using the `\ifvmode`, `\ifhmode`, `\ifmmode` and `\ifinner` `\if`-tests. The first three of these tests return "true" if the macro is invoked in a vertical mode, a horizontal mode, or a math mode, respectively. The last test returns "true" if TeX is in one of the inner modes: in the restricted horizontal mode, in the internal vertical mode, or in the ordinary math mode.

EXAMPLE 8.1 ▷ The following macro displays the current mode of operation:

```
\def\tellmode{\message{We are in}
  \ifhmode
    \ifinner\message{Restricted Horizontal Mode}
      \else\message{Horizontal Mode}\fi
    \else
  \ifvmode
    \ifinner\message{Internal Vertical Mode}
      \else\message{Vertical Mode}\fi
    \else
    \ifinner\message{Math Mode}
      \else\message{Display Math Mode}\fi
    \fi
  \fi}
```

8.2 Rules

TOPICS ▷ \hrule, \vrule, \strut, depth, height, width

The \vrule and \hrule commands typeset vertical and horizontal solid-black lines or rectangles (*rules*). Both commands allow optional parameters that specify the dimensions of the rule. These two commands have similar syntax and produce similar results. \hrule may be used only in vertical modes to produce (usually) horizontal rules; \vrule is used in horizontal modes to produce (usually) vertical rules.

There are three possible dimensional parameters. The ⟨width⟩ parameter sets the (horizontal) width of a rule, ⟨height⟩ sets the height of the rule *above the base line*, and ⟨depth⟩ sets the height of the rule *below the base line*.

SYNTAX ▷ \hrule [width ⟨width⟩] [depth ⟨depth⟩] [height ⟨height⟩]
▷ \vrule [width ⟨width⟩] [depth ⟨depth⟩] [height ⟨height⟩]

The ⟨width⟩, ⟨depth⟩, and ⟨height⟩ parameters can be specified in any order; if the same parameter is specified more than once, the *last* specification is used.

EXAMPLE 8.2 ▷ Typing

```
\hrule \vskip 3pt
\hrule height 0.1pt \vskip 3pt
\hrule height 1.0pt width 3in \vskip 3pt
\hrule height 2pt  width 1in \vskip 3pt
```

results in

Here the `\vskip` commands prevent rules from colliding.

Horizontal rules appear in vertical mode. If their `width` is omitted, the default is the width of the page. Inside boxes, described in the next section, the width defaults to the width of the box. Unspecified `height` and `depth` default to `0.4pt` and `0pt`.

EXAMPLE 8.3 ▷ Typing

```
A \vrule\ B \par
A \vrule height 3mm\ B \par
A \vrule width 3mm\ B \par
A \vrule depth 3mm\ B \par
A \vrule depth 3mm height 3mm\ B \par
```

produces

A | B
A | B
A ■ B
A | B
A | B

If `width` field of `\vrule` is not specified, TEX assumes the width to be `0.4pt`. Missing `height` or `depth` fields are filled in order to make the rule as high as the current line of text. For instance, if the previous example were typeset in one long line, instead of five short ones, all rules would become `3mm` high and `3mm` deep.

Both `\depth` and `\height` may be negative. To create a 5pt-wide `\vrule` with the lower end 3 points above the base line and with the upper end 6 points above the base line (like this: ▪), type

```
\vrule width 5pt height 6pt depth -3pt
```

The ____ `\vrule` has `width 1cm`, `height -2pt`, and `depth 1cm`.

➡ As the above examples show, a `\vrule` might look like a horizontal rule. The difference between `\hrule` and `\vrule` is not their shapes, but the mode in which they are used and the treatment of the defaults.

A rule becomes invisible if the sum of the height and depth is non-positive or if its width is non-positive. Invisible rules are useful if there are lines in a paragraph or a table with different heights (for instance, if you use large symbols, equations with fractions or boxes inside a paragraph) and you wish to make all the lines of the same height. Inserting a large invisible rule in a line will make the line as high as the rule. For example, an invisible rule `\vrule width -1pt depth 7pt height 7pt` was inserted in this paragraph just before the words "For example". This made the line higher.

The `\strut` command generates an invisible rule that is `8.5pt` high and `3.5pt` deep. It is especially useful in tables (§9.3-§9.4) where it is important to make all rows of the same height.

8.3 Boxes

TOPICS ▷ `\hbox`, `\vbox`, `\vtop`, `\vcenter`

As was observed in §5.4, typesetting with TeX can be viewed as putting together metal plates (or *boxes*) with glue. There are different kinds of boxes; the aim of TeX is to start with the smaller boxes, such as characters, and eventually glue them together into the biggest possible box, a page.

There are essentially two types of boxes:

- *vertical boxes* have their contents arranged vertically; these boxes are connected or "glued" together horizontally
- *horizontal boxes* have their contents arranged horizontally; these boxes are connected or "glued" together vertically

Boxes that may occur during typesetting include:

- single character boxes.
- rule boxes, both vertical and horizontal.
- `\hbox` word boxes built from character boxes to prevent line breaks.
- line boxes built from word boxes when all the line breaks are found. These boxes are usually not explicitly entered by the user; instead TeX forms them automatically in the process of breaking paragraphs into lines. Forming these boxes is the main function of the horizontal mode.
- page boxes built from line boxes. Again, these boxes are not explicitly entered by the user; instead TeX forms them automatically in the process of breaking

text into pages. Forming these boxes is the main function of the vertical mode.

- picture boxes that either contain graphics or are used to reserve space for graphics.

Each box is described by three main parameters: its width, height (height over the base line), and depth (height below the base line).

When TeX deals with characters, it is unaware of the actual shapes of the characters being typeset (this is important only for the device drivers); TeX knows only the dimensions of the box that surrounds a character and a few other things, such as kerning tables and the list of the characters that form ligatures. Likewise, once TeX builds a box, it forgets its contents. Only the dimensions of the box will be used to build bigger boxes.

In most cases, the typesetting is a two-step procedure: lines are built from characters in the horizontal mode and then pages are built from lines in the vertical mode. However, in more advanced examples, you may see vertical building done inside a line of text (internal vertical mode) or horizontal building inside a vertical mode (restricted horizontal mode). The next example illustrates this.

EXAMPLE 8.4 ▷ Typing

```
\font\eightrm=mvtmsr scaled 800
\font\eightti=mvtmsi scaled 800 slant 220
There was a sign on the fence:
\vrule\
\vbox{\baselineskip 8pt \eightrm
\hbox{Do not walk}\par \hbox{on the grass!}} \ \vrule

And there was a graffito underneath:
\vrule\
\vtop{\baselineskip 8pt \eightti
\hbox{Why not walk}\par \hbox{on the grass?}} \ \vrule
```

produces

| Do not walk |
There was a sign on the fence: | on the grass! |
And there was a graffito underneath: | *Why not walk* |
| *on the grass?* |

The `\vbox` command creates a vertically arranged box that contains two lines of text: "`Do not walk`" and "`on the grass`". Observe that inside the `\vbox`, TeX behaves as if it were typesetting an entire page: the `\par` command denotes the end of a paragraph and the interline spacing can be locally changed by `\baselineskip`. The `\vtop` command used to typeset the second sentence is similar to `\vbox` except that the box it constructs is treated differently: it is aligned to the top instead of being aligned to the bottom. The `\hbox` command places its contents in a horizontal box; the width of the box is the natural width of its contents. In the example `\hbox` is necessary since otherwise the lines that contain "`Do not walk`" and "`on the grass`" would be made as wide as the page; then the encompassing `\vbox` would also be as wide as the page and could not be merged inside another paragraph. Finally, the `\vrule` commands produce vertical bars around signs.

SYNTAX ▷ `\hbox [ to` ⟨dimen⟩ `] [ spread` ⟨dimen⟩ `] {...}`.

The dimensions of an `\hbox` depends on the arguments that may be passed to `\hbox`. You may use `\hbox` without any dimension arguments, and TEX will typeset the box with its natural width. If a `to` ⟨dimen⟩ is specified, then the width of the hbox will be equal to ⟨dimen⟩. If a `spread` ⟨dimen⟩ is specified, then the width of the box will be its natural width plus ⟨dimen⟩.

EXAMPLE 8.5 ▷ Typing

```
\leftline{\hbox spread - 4pt{Mary had a little lamb.}}
\leftline{\hbox spread - 2pt{Mary had a little lamb.}}
\leftline{\hbox              {Mary had a little lamb.}}
\leftline{\hbox spread   2pt{Mary had a little lamb.}}
\leftline{\hbox spread   4pt{Mary had a little lamb.}}
```

causes the following display, as well as some complaints from TEX, forced to form overfull and underfull lines.

Mary had a little lamb.
Mary had a little lamb.
Mary had a little lamb.
Mary had a little lamb.
Mary had a little lamb.

Inside a horizontal box, the text is evenly distributed. Horizontal skips and glues may be used to format the text.

➡ The `\line` command is internally defined as a macro:

```
\def\line{\hbox to\hsize}
```

Recall that `\hsize` is the horizontal size of the printable page.

The features of horizontal boxes include:

- The text inside an `\hbox` is kept together: a box cannot be split across lines. Thus, you can use `\hbox` to suppress hyphenation (see §3.10).
- The contents of an `\hbox` can be copied (see `\copy` below) or moved (see `\kern`, `\raise`, `\lower`) as a whole.
- Inside an `\hbox` TeX operates in restricted horizontal mode.
- An `\hbox` appearing in math mode may include text typeset in any font.

The `\vbox` command constructs vertical boxes. It is used like `\hbox`; optional parameters `to` and `spread` may be used. `\vbox` inserted in a paragraph has its bottom on the base line.

The `\vtop` command creates a box that looks like one created by `\vbox`. However, when a `\vtop` box is merged into a paragraph, its top line aligns with the paragraph line.

EXAMPLE 8.6 ▷

```
{\leftline{\raggedright\vtop{\hsize = 2in
The text that will be typeset within this box
will be referenced at its top.}
\qquad % put some space between boxes
\vbox{\hsize = 2in
And this box will have its reference located near
or at its bottom.}
} }
```

produces

	And this box will have its reference located near or at its bottom.
The text that will be typeset within this box will be referenced at its top.	

Remember that the `\hsize` command normally sets the width of a page. However, in the previous example it is changed to set the width of the text in the vertical box. Thus, while within the box, TeX justifies the text in a two-inch column; once the box is completed, `\hsize` is automatically restored and the text is justified to the page width.

There are two boxes in the above example, one with its reference at the top of the box and the other with its reference at the bottom. The `\leftline` command inserts these boxes by placing their reference points at the baseline of the new line being typeset.

The `\vcenter` command also creates a vertical box identical to one created by `\vbox`, but a `\vcenter` box will be centered on the current line. `\vcenter` may be used only in math mode. `\vcenter` is internally used to position fractions and matrices. To use `\vcenter` in the text mode, you may need to temporarily enter the math mode.

8.4 Box Variables – Operations on Boxes

TOPICS ▷ `\setbox`, `\box`, `\copy`, `\wd`, `\ht`, `\dp`, `\newbox`

Boxes do not necessarily need to be processed immediately. Instead, you can use up to 256 special *box variables*, which hold boxes. Box variables can be accessed either by their number (0 through 255) or by their name.

The `\newbox` command allocates a new box variable.

SYNTAX ▷ `\newbox` ⟨box variable⟩

Box variables are local to the groups in which they are declared. In effect, more than 256 boxes may be used.

The `\setbox` command sets the value of a box variable.

SYNTAX ▷ `\setbox` ⟨box variable or number⟩ = ⟨box expression⟩

where the ⟨box variable⟩ argument denotes a box variable or a number with a range of 0 to 255, and ⟨box expression⟩ is a box created by `\hbox`, `\vbox`, or for that matter, any of the other box commands.

EXAMPLE 8.7 ▷ Typing

```
\newbox\testbox
\setbox\testbox =
   \hbox{(\vrule $ {123 \over 456} \heartsuit)$}
```

allocates and sets a box variable `\testbox`. Typing

```
\setbox 7 = \hbox{(\vrule $ {123 \over 456} \heartsuit)$}
```

sets the seventh box variable.

The `\box` and `\copy` commands insert the contents of a box variable into the document. The `\box` command destroys the contents of the box variable, while `\copy` keeps the contents. Two commands are needed because box variables are often expensive in terms of storage; it is best to clear them as soon as they are no longer needed.

SYNTAX ▷ `\box` ⟨box variable or number⟩
▷ `\copy` ⟨box variable or number⟩

EXAMPLE 8.8 ▷ Typing

```
\setbox 7 = \hbox{(\vrule $ {123 \over 456} \heartsuit)$}
boxes: \copy7 \copy7 \copy7 \copy7 \copy7 \box7 \box7
```

produces

boxes: (|$\frac{123}{456}\heartsuit$)(|$\frac{123}{456}\heartsuit$)(|$\frac{123}{456}\heartsuit$)(|$\frac{123}{456}\heartsuit$)(|$\frac{123}{456}\heartsuit$)(|$\frac{123}{456}\heartsuit$)

Observe that only six boxes appear since the first occurrence of the `\box7` command emptied the box after typesetting it.

The `\ifvoid`, `\ifhbox`, and `\ifvbox` commands test whether a box is void (uninitialized or destroyed), is a horizontal box, or is a vertical box (see §6.9).

EXAMPLE 8.9 ▷ Typing

```
\newbox\mybox
\setbox\mybox=\hbox{(\vrule ${123 \over 456}\heartsuit)$}
\message{This is a }
\ifvoid\mybox\message{void}\else\message{non-empty}
  \ifhbox\mybox\message{horizontal}%
      \else\message{vertical}\fi
  \message{ } \fi
  \message{ box.}
```

causes TeX to print

```
This is a non-empty horizontal box.
```

on the screen.

Boxes, like rules, have three dimensions: width, height, and depth. The `\wd`, `\ht`, and `\dp` commands retrieve these dimensions.

SYNTAX ▷ `\wd` ⟨box variable or number⟩
▷ `\ht` ⟨box variable or number⟩
▷ `\dp` ⟨box variable or number⟩

EXAMPLE 8.10 ▷ Typing

```
\newbox\NYbox
\setbox\NYbox=\hbox{$\hbox{\sf I}%
  \heartsuit\over\hbox{\sf NY}$}
Box “\copy\NYbox” has width \the\wd\NYbox, height
\the\ht\NYbox, and depth \the\dp\NYbox.
```

results in

> Box “$\frac{\mathsf{I}\heartsuit}{\mathsf{NY}}$” has width 16.15005pt, height 10.9518pt, and depth 5.0444pt.

8.5 Moving Boxes – Kerning

TOPICS ▷ `\lower`, `\raise`, `\moveleft`, `\moveright`, `\kern`

In horizontal modes, `\lower` and `\raise` will lower and raise the position of a horizontal box.

SYNTAX ▷ `\lower` ⟨dimen⟩ ⟨box variable or number⟩
▷ `\raise` ⟨dimen⟩ ⟨box variable or number⟩

Any ⟨dimen⟩ units may be used. However, if `\raise` and `\lower` are used to position subscripts or superscripts in the text mode, it is best to use the font-dependent unit `ex` (the height of the letter 'X').

EXAMPLE 8.11 ▷ This example illustrates the formation of the “™” (trademark) symbol, which is omitted in traditional TeX typefaces. This is irrelevant for VTeX users, since VTeX fonts include the proper trademark symbol.

For example,

```
\font\text = mvmelr \text
\font\smallletters = mvmelr scaled 500
\def\tm {\raise .7ex \hbox{\smallletters TM}}
\def\TM{\char154}
\TeX\tm\ users form trademark while
V\TeX\TM\ users use the proper character.
```

results in

> TeX™ users form trademark while VTeX™ users use the proper character.

Similarly, in vertical mode, the `\moveright` and `\moveleft` commands move a box, usually a line of text, right or left.

SYNTAX ▷ `\moveleft` ⟨dimen⟩ ⟨box⟩
▷ `\moveright` ⟨dimen⟩ ⟨box⟩

EXAMPLE 8.12 ▷ Typing

```
\def\abit{0.3in}      \def\alot{0.6in}
\centerline{\bf This is the center.}
\moveleft  \abit \vbox{\centerline{The old government
       of Banana Islands was a bit left of center.}}
\moveright \alot \vbox{\centerline{The new government
       of Banana Islands is a lot to the right.}}
```

produces

> **This is the center.**
> The old government of Banana Islands was a bit left of center.
> The new government of Banana Islands is a lot to the right.

The `\raise`, `\lower`, `\moveleft`, and `\moveright` commands may not be applied to elementary boxes, that is, characters and rules. The ⟨box⟩ argument must be an explicitly specified box, created by `\hbox`, `\copy`, or similar commands. However, in the last example, the `\vbox` command is not needed, since the `\centerline` macro itself starts with a box command.

The `\moveleft` and `\moveright` commands should not be used to adjust the inter-character spacing. Instead, use the `\kern` command.

SYNTAX ▷ `\kern` ⟨dimen⟩

where ⟨dimen⟩ specifies the amount of horizontal motion. If ⟨dimen⟩ is positive, the motion is to the right; otherwise it is to the left.

EXAMPLE 8.13 ▷ The definition of the `\TeX` macro is the following:

```
\def\TeX{T\kern-.1667em\lower.5ex\hbox{E}\kern-.125emX}
```

The dimensions in kern commands can be expressed in many units besides em's and ex's. It is, however, best to use only these units, since they are font and resolution independent. This allows TEX's logo to be reasonably printed in widely different fonts:

TEX . TEX . TEX . TEX . TEX . TEX

Standard TEX also uses kerning to typeset the copyright symbol ©. The symbol is built from a small 'c' and a circle appearing in one of the math fonts. VTEX fonts include proper `\copyright` (©) and `\registered` (®) symbols as characters 152 and 153.

8.6 Drop-Case Letters

The examples in this section combine the use of boxes with macros, variables, and paragraph shaping commands. In a way, they bring together all of the TEX components. The first example we will consider is the "drop-case" paragraph formatting, an essential component of old books and still the preferred way for printing classical works.

The problem of typesetting a drop-case paragraph is in performing the calculations needed to determine the paragraph shape, depending on the size of the main font and the size of the drop-case letter. In the example below we will set the text using the 10-point `mvpalr` and the drop-case letter in the 40-point `mvcloi`. A fairly lengthy macro computes the `\hangindent` and `\hangafter` values. This macro was contributed by Niu Can.

```
\def\dropcase#1#2{{\noindent
\setbox0\hbox{\font\big=mvcloi scaled 4000\big #1}%
\setbox1\hbox{#2}%
\setbox2\hbox{()}%
\count0=\ht0\advance\count0 by\dp0
\count1=\baselineskip
\advance\count0 by -\ht1\advance\count0 by \ht2
\dimen1=0.5ex\advance\count0 by\dimen1
\divide\count0 by \count1
\advance\count0 by 1\dimen0\wd0
\advance\dimen0 by .5em\dimen1=\ht0\advance\dimen1 by -\ht1
\global\hangindent\dimen0 \global\hangafter-\count0
\hskip-\dimen0\setbox0
\hbox to \dimen0{\raise-\dimen1\box0\hss}%
\dp0=0in\ht0=0in\box0}#2\par}
```

Then, typing

```
\dropcase{Q}{uite often...
```

results in

Quite often this type of layout was used in rare first editions. Some of the most expensive rarities sold at Christie's in recent years featured the drop-case technology; it was mentioned by critics that use of drop-case raised the price by as much as 17.3%. A few recent record setters include Shakespeare's "Night at Waterloo", Machiavelli's "On pity", Newton's "On breeding apples for softness", and Wagner's biography of Mendelsohn. Some critics, however, regard "Night at Waterloo" as actually written by Bacon.

The reader will notice that one can change the macro's declaration to

```
\def\dropcase#1#2\par{...
```

to avoid having to surround the drop-case letter in braces.

8.7 Shipping out

TOPICS ▷ `\shipout`, `\output`

TEX constructs boxes even if box commands do not explicitly appear in the document. TEX first forms line boxes and then merges them into a page box. The page box is then decorated with headers and footers and written to the `.DVI` file. The actual writing is done with the `\shipout` command.

SYNTAX ▷ `\shipout`⟨box⟩

`\shipout` works immediately, regardless of the page-breaking mechanism. For example, if your document contains

```
Before
\shipout\hbox{and}
After
\end
```

The output will consists of two pages; first will contain the word 'and' (and no page number), while the second will have 'Before', 'After' and will be numbered 1. `\shipout` does not do anything beyond shipping a box out; it does not increment the page number or append it to the page as part of the footer.

Normally, the document will not call `\shipout` explicitly. Instead, when a page body has been determined, TEX uses the `\output` command; `\output`, in turn, completes the page and only then calls `\shipout`. The body of the page can be found in the `\box255` register.

SYNTAX ▷ `\output{`⟨output routine⟩`}`

where ⟨output routine⟩ is a macro responsible for completion of the page and shipping it out. A simplest meaningful output routine would be

```
\def\simpleoutput{\shipout{\box255}
  \advancepageno }
\output{\simpleoutput}
```

An actual output routine used in TeX is substantially more complicated. While the default output routine in TeX is sufficient in most cases, one would write special-purpose output routines to achieve unusual formatting. The most common example is the double-column page format as used in indexes and dictionaries.

8.8 Filling. Leaders

TOPICS ▷ `\leaders`, `\hrulefill`, `\dotfill`, `\upbracefill`, `\downbracefill`
▷ `\rightarrowfill`, `\leftarrowfill`, `\cleaders`, `\xleaders`

In tables of contents and forms it is frequently necessary to fill a horizontal space with a rule or dots. Instead of just glue (see §5.4), horizontal spaces can be filled with any boxes.

SYNTAX ▷ `\leaders` <horizontal box or rule> ⟨glue⟩

This replaces ⟨glue⟩ by repeated occurrences of ⟨horizontal box or rule⟩.

EXAMPLE 8.14 ▷ Define `\dotbox` to be a `1em` wide box containing a dot:

```
\def\dotbox{\hbox to 1em{\hss.\hss}}
```

Typing

```
1. Introduction\leaders\dotbox\hfill 1  \par
2. Credits and Acknowledgements
        \leaders\dotbox\hfill 12 \par
3. The Real Story\leaders\dotbox\hfill 34 \par
4. Conclusion\leaders\dotbox\hfill 187
```

results in

1. Introduction 1
2. Credits and Acknowledgements 12
3. The Real Story 34
4. Conclusion 187

EXAMPLE 8.15 ▷ Using

```
\def\dotbox{\hbox to 1em{\hss.\hss.\hss}}
```

in the previous example yields

1. Introduction	1
2. Credits and Acknowledgements	12
3. The Real Story	34
4. Conclusion	187

Notice that the `\leaders` command always aligns dots (unlike `\cleaders` and `\xleaders` described below).

A horizontal rule may be used instead of dots.

EXAMPLE 8.16 ▷ Typing

```
\leftline{{\bf\underbar{Information Form}}}
First Name: \leaders\hrule\hskip 2.5in \hss\par
Last Name: \leaders\hrule\hskip 2.5in \hss\par
SS\#: \leaders\hrule\hskip 1.5in \hss\par
Address: \leaders\hrule\hskip 3.5in \hss\par
City, State, Zip: \leaders\hrule\hskip 3.0in \hss\par
```

produces

Information Form
First Name: ____________________________
Last Name: ____________________________
SS#: ________________
Address: ______________________________________
City, State, Zip: _________________________________

The `\hrulefill` command is equivalent to `\leaders\hrule\hfill`.

EXAMPLE 8.17 ▷ Typing

```
1. Introduction\hrulefill\ 1  \par
2. Credits and Acknowledgements\hrulefill\ 12 \par
3. The Real Story\hrulefill\ 34 \par
4. Conclusion\hrulefill\ 187 \par
```

produces

1. Introduction____	1
2. Credits and Acknowledgements__________________________	12
3. The Real Story_____________________________________	34
4. Conclusion____	187

The alignment of the `\dotbox` boxes in the above examples is a consequence of the strategy used by `\leaders`: first the entire line is filled with boxes; next the boxes that fit into the glue window are actually displayed. Two different packing strategies correspond to two more types of leaders: centered leaders `\cleaders` and expanded leaders `\xleaders`. The `\cleaders` command packs the filling boxes tightly to each other and distributes the remaining space on the sides; the `\xleaders` command evenly distributes the available space.

EXAMPLE 8.18 ▷ Typing

```
\def\cfill{\cleaders\hbox to 1em{\hss.\hss}\hfill}
1. Introduction\cfill                         1 \par
2. Credits and Acknowledgements\cfill        12 \par
3. The Real Story\cfill                      34 \par
4. Conclusion\cfill                         187 \par
```

results in

1. Introduction	1
2. Credits and Acknowledgements	12
3. The Real Story	34
4. Conclusion	187

Using `\xleaders` in the above example produces

1. Introduction	1
2. Credits and Acknowledgements	12
3. The Real Story	34
4. Conclusion	187

The difference between `\cleaders` and `\xleaders` becomes more visible when the dots are spread more thinly.

All three types of leaders produce identical results when used with an `\hrule`.

A few filling commands are predefined in TeX. They are:

- `\dotfill` fills with dots
- `\upbracefill` fills with a brace facing up
- `\downbracefill` fills with a brace facing down
- `\rightarrowfill` fills with a right arrow
- `\leftarrowfill` fills with a left arrow

`\dotfill` is most commonly used for building tables of contents. The last four of these commands are used in math mode as part of the definition of `\overbrace`, `\underbrace`, `\uprightarrow`, and `\upleftarrow` (see §4.11).

`\leaders`, `\cleaders`, and `\xleaders` may also be used in vertical mode. In this case, vertical boxes and glue should be used instead of horizontal ones.

8.9 Gray Rules and Patterns

TOPICS ▷ `\gray`, `\grayl`, `\grayr`, `\graypattern`

✔ This section is applicable to VTEX only.

Gray rules are often used for special visual effects. To typeset a gray rule, use the `\gray` prefix before either the `\vrule` or `\hrule` command.

SYNTAX ▷ `\gray \hrule` ⟨dimensions⟩
▷ `\gray \vrule` ⟨dimensions⟩

EXAMPLE 8.19 ▷ Typing

```
\centerline{\gray \vrule height 2mm width 2in}
```

produces

The `\grayl` and `\grayr` prefix commands employ different gray patterns. If you used these commands in the previous example instead of `\gray`, you would obtain for `\grayl`

and for `\grayr`

Gray rules are built from the *pattern* rules: smaller rectangles which follow the same pixel pattern. The `\graypattern` command allows you to design your own gray patterns. `\graypattern` changes the pattern used by `\gray`; it has no effect on `\grayl` and `\grayr`.

SYNTAX ▹ `\graypattern{`⟨pattern specification⟩`}`

where the ⟨pattern specification⟩ describes the arrangements of the pixels in the pattern rectangle. You may specify up to 32 rows and 8, 16, 24, or 32 columns in the pattern. ⟨pattern specification⟩ contains the specifications for individual pixel rows, separated by the slash character `/`. Each row is specified by giving the hexadecimal number that describes the pattern. The hexadecimal digits `0`, `1`, `2`, `3`, `4`, `5`, `6`, `7`, `8`, `9`, `A`, `B`, `C`, `D`, `E`, and `F` code patterns of four pixels as shown below:

0	○ ○ ○ ○	8	● ○ ○ ○
1	○ ○ ○ ●	9	● ○ ○ ●
2	○ ○ ● ○	A	● ○ ● ○
3	○ ○ ● ●	B	● ○ ● ●
4	○ ● ○ ○	C	● ● ○ ○
5	○ ● ○ ●	D	● ● ○ ●
6	○ ● ● ○	E	● ● ● ○
7	○ ● ● ●	F	● ● ● ●

EXAMPLE 8.20 ▹ To specify the pattern

● ● ○ ○ ○ ○ ○ ●
○ ○ ○ ○ ○ ○ ● ○
○ ○ ○ ○ ○ ● ○ ○
● ● ● ● ● ○ ○ ●

you should enter

```
\graypattern{C1/02/04/F9}
```

Now, typing

```
\centerline{\gray \vrule height 2mm width 2in}
```

produces

In designing a gray pattern, keep in mind that a pattern containing more than 10%-15% of black pixels will appear black on most printers. The predefined patterns `\gray`, `\grayl`, and `\grayr` contain very few black pixels. The `\grayl` pattern, for instance, is `0001/0010/0100/1000` or, graphically,

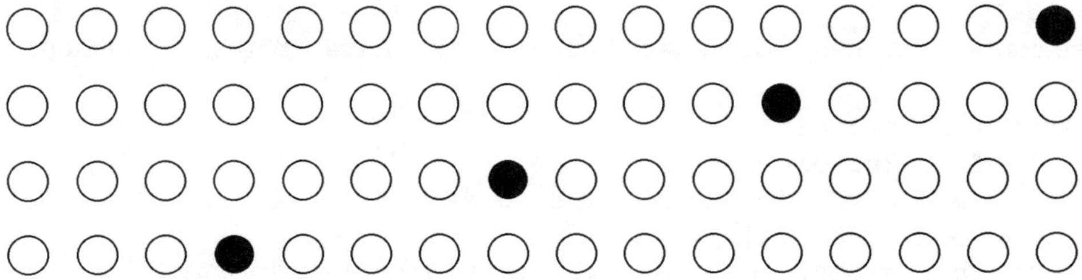

The `\grayr` pattern is defined as `1000/0100/0010/0001` and the `\gray` pattern is defined as `01/00/10/00`.

➡ Gray patterns (like character fill patterns) are inherently device-dependent (alas). This means that you may need to design somewhat different patterns if you are to print your document at high resolution. The predefined patterns shown here are only good at 300 dots per inch (LaserJet) or lower. When this book was first printed at 600 dpi, some of the patterns became black; some looked uneven. VTEX device drivers can correct the problem with the fill patterns by using different tables for different devices; however they cannot correct gray rule patterns.

8.10 Broken Underline. Strikeout. Redline

Using a combination of boxes with filling and gray shades, one can design macros to do broken underlining, strikeout and, redline. The approach is to place the affected text into a box, then build a box of the same width with a correct rule pattern, and, finally, superimpose the boxes. We will start with underlining:

```
\def\brunderbar#1{%
  \def\brokenrule
    {\vrule width 0.20em height -0.2ex depth 0.3ex}%
  \setbox1=\hbox{#1}%
  \setbox2=\hbox to 0.5em{\brokenrule\hss\brokenrule}%
  \setbox3=\hbox to \wd1{\leaders\box2\hfill}%
  \wd3=0pt
  \hbox{\box3\box1}%
  }
```

Now, typing

```
The following will be \brunderbar{underlined} with a
broken (as opposed to solid) rule.
```

results in

The following will be underlined with a broken (as opposed to solid) rule.

The strikeout effect can be achieved by simply changing the dimensions of `\brokenrule`:

```
\def\strikeout#1{%
  \def\brokenrule
    {\vrule width 0.20em height 0.4ex depth -0.3ex}%
  \setbox1=\hbox{#1}%
  \setbox2=\hbox to 0.5em{\brokenrule\hss\brokenrule}%
  \setbox3=\hbox to \wd1{\leaders\box2\hfill}%
  \wd3=0pt
  \hbox{\box3\box1}%
  }
```

Now, typing

```
Current assets of the corporation are estimated
at 10 \strikeout{million} dollars.
```

results in

Current assets of the corporation are estimated at 10 ~~million~~ dollars.

✔ Finally, redlining can be obtained with the use of gray rules. `\brokenrule` becomes

```
\def\grayp{\special{M01/00/10/00}}
\def\brokenrule{\grayp
   \vrule width 0.20em height 1.3ex depth -0.2ex}%
```

Assuming a definition of `\redline` modelled upon those of `\strikeout` and `\brunderbar`, typing

```
The leaser agrees to vacate the premises
\redline{as soon as convenient}.
```

results in

The leaser agrees to vacate the premises as soon as convenient.

The dimensions of the rules in the above examples can be changed; it is, however, best to express them in the `ex` units.

Chapter 9

Tabulation & Tables

9.1 Tabulation — Evenly Spaced Tabs

TOPICS ▷ `\settabs`, `\columns`, `\+`, `\cr`, `\tabalign`

The simplest way to set tabs is to set them at even intervals.

SYNTAX ▷ `\settabs` ⟨number⟩ `\columns`

sets ⟨number⟩ tabs at even intervals.

EXAMPLE 9.1 ▷ `\settabs 5 \columns` divides the page into five columns, all of the same width.

Tabulated lines should satisfy the following syntax rules:

SYNTAX ▷ `\+` ⟨text-1⟩ `&` ⟨text-2⟩ `&` ... ⟨text-N⟩ `\cr`

Here `\+` starts a tabulated line, `\cr` ends it, ⟨text-1⟩ ... ⟨text-N⟩ contain text to be placed in columns 1 through N, and `&` is used to separate columns.

EXAMPLE 9.2 ▷ Typing

```
\settabs 5 \columns
\+ Company & Quarter 1 & Quarter 2
           & Quarter 3 & Quarter 4 \cr
\smallskip
\+ Jones \& Sons & 12.32 & 13.34 & 14.32 & 15.11 \cr
\+ UnderDog Inc  & 33.33 & 35.43 & 36.35 & 45.03 \cr
\+ \EEFlogo      & 23.33 & 24.34 & 26.12 &  0.72 \cr
```

produces

Company	Quarter 1	Quarter 2	Quarter 3	Quarter 4
Jones & Sons	12.32	13.34	14.32	15.11
UnderDog Inc	33.33	35.43	36.35	45.03
$\mathcal{EEF}$	23.33	24.34	26.12	0.72

The entries in a table constructed with `\settabs` are left-justified. Glues such as `\hfill` should be used for centering or right-justification. For example, inserting `\hfill`'s in columns 2,3,4 and 5 of the previous example, as in

```
\+ Jones \& Sons &\hfill 12.32 &\hfill 13.34
                 &\hfill 14.32 &\hfill 15.11\cr
```

causes the table to become

Company	Quarter 1	Quarter 2	Quarter 3	Quarter 4
Jones & Sons	12.32	13.34	14.32	15.11
UnderDog Inc	33.33	35.43	36.35	45.03
$\mathcal{EEF}$	23.33	24.34	26.12	0.72

If the text does not fit into a column, it extends to the next one. In this case TEX does not generate any error messages.

EXAMPLE 9.3 ▷ Replacing $\mathcal{EEF}$ with "**Exotic and Edible Foods**, *the accepted leader of the industry*" in the last example results in

Company	Quarter 1	Quarter 2	Quarter 3	Quarter 4
Jones & Sons	12.32	13.34	14.32	15.11
UnderDog Inc	33.33	35.43	36.35	45.03
Exotic and Edible Foods, *the accepted leader of the industry*	23.33	24.34	26.12	0.72

To typeset this table you should use variable-width tabs, described in the next section.

The following rules describe the use of `\settabs`. They also apply to variable-width tabs.

1. If you specify `\column N`, the width of the page is divided onto N columns of equal size. In each tabulated line, text before `&` is placed into the first column, text after the first `&` and before the second `&` is placed in the second column, and so on.
2. Extra `&` characters are ignored; no error message is printed.
3. Tabulated lines do not have to correspond to input lines in the source; however, placing tabulated lines on separate source lines improves the clarity of the source.
4. Spaces after `&` and `\+` are ignored.
5. Each entry is treated as a group. For example, font specifications in an entry affect only that entry.
6. It is not necessary to specify all the columns. If `\cr` is encountered before all columns are specified, the remaining columns will be blank.
7. Tabbed lines may be intermixed with ordinary text.
8. Defining macros for abbreviation of special formatting of entries of the table will lead to a substantial saving in typing.
9. A table built with `\settabs` and `\+` may not be used inside a macro, since `\+` is declared as an `\outer` macro to safeguard against potential errors. The `\tabalign` command is identical to `\+`, except that it can be used anywhere. For instance, in the source file for this book, the examples for this section were actually typeset with `\tabalign` instead of `\+` since they were further placed inside a box.

9.2 Tabulation – Arbitrary Tabs

TOPICS ▷ `\settabs`, `\+`, `\cr`, `\tabalign`

Instead of specifying only the number of columns, the user may define the size of each column.

SYNTAX ▷ `\settabs\+`⟨pattern line⟩ `\cr`

If `\settabs` is immediately followed by `\+` and a ⟨pattern line⟩, the widths of the entries in the ⟨pattern line⟩ are used to set the tabs. ⟨pattern line⟩ itself is not typeset.

EXAMPLE 9.4 ▷ The text

```
\settabs\+MMMMMMMMMMMMMMMM&MMMMMM&MMMMMM&MMMMMM&MMMMMM\cr
\+ Company & Quarter 1 & Quarter 2
          & Quarter 3 & Quarter 4 &\cr
\smallskip
\+ Jones \& Sons & 12.32 & 13.34 & 14.32 & 15.11 &\cr
\+ UnderDog Inc  & 33.33 & 35.43 & 36.35 & 45.03 &\cr
\+ \EEFlogo      & 23.33 & 24.34 & 26.12 &  0.72 &\cr
```

is typeset as

Company	Quarter 1	Quarter 2	Quarter 3	Quarter 4
Jones & Sons	12.32	13.34	14.32	15.11
UnderDog Inc	33.33	35.43	36.35	45.03
$\mathcal{EEF}$	23.33	24.34	26.12	0.72

The pattern line

```
MMMMMMMMMMMMMMMM&MMMMMM&MMMMMM&MMMMMM&MMMMMM\cr
```

is used to compute the actual settings of tabs. It is important to remember that in computing the exact tab positions, TEX uses the widths of the characters of the pattern line. These widths are taken from the font that was current when the `\settabs` command was invoked. Since most of the fonts are proportional, it is preferable to use only one pattern character in the pattern line.

You can also specify the tab setting exactly, by using the `\hskip` command inside the pattern line.

EXAMPLE 9.5 ▷ In the previous example, using the pattern line

```
\settabs\+\hskip 1.5in&\hskip 1.7cm&\hskip
 1.7cm&\hskip 1.7cm&\hskip 1.7cm&\cr
```

results in the table

Company	Quarter 1	Quarter 2	Quarter 3	Quarter 4
Jones & Sons	12.32	13.34	14.32	15.11
UnderDog Inc	33.33	35.43	36.35	45.03
$\mathcal{EEF}$	23.33	24.34	26.12	0.72

9.3 Alignment Templates

TOPICS ▷ `\halign`,`\cr`, `\tabskip`, `to`, `spread`

Using the tabulation command **`\settabs`** requires an a priori knowledge of the width of the columns. In many cases it is easier to supply only a description of the organization of the table and the actual information to be arranged in columns, and then let TEX do the computations of widths. This may be done using the **`\halign`** (horizontal align) command. The **`\halign`** command resembles the pattern line form of `\settabs`. The command is followed by a template ending with a `\cr` and then by the rows of the table.

SYNTAX ▷ `\halign{` ⟨template⟩ `\cr` ⟨text⟩ `\cr` ⟨text⟩ `\cr` ... `\cr }`

The ⟨template⟩ may contain

- the character & which separates the columns in the ⟨template⟩
- the character # which is replaced by actual text
- other characters or commands that are copied literally to each line of the table

⟨text⟩ contains the substitution text. ⟨text⟩ consists of rows of text, each row ending with a `\cr`. Inside a row, the substitution items are separated by the alignment tab &.

EXAMPLE 9.6 ▷ Text

```
\halign{ \quad #\hfil & \hfil#
               & \hfil# & \hfil# & \hfil# \cr
 Company & Quarter 1 & Quarter 2
         & Quarter 3 & Quarter 4 \cr
 Jones \& Sons & 12.32 & 13.34 & 14.32 & 15.11 &\cr
 UnderDog Inc  & 33.33 & 35.43 & 36.35 & 45.03 &\cr
 \EEFlogo      & 23.33 & 24.34 & 26.12 &  0.72 &\cr
 }
```

will be typeset as

Company	Quarter 1	Quarter 2	Quarter 3	Quarter 4
Jones & Sons	12.32	13.34	14.32	15.11
UnderDog Inc	33.33	35.43	36.35	45.03
$\mathcal{EEF}$	23.33	24.34	26.12	0.72

Here the ⟨template⟩ contains five entries, corresponding to five columns. The first column is specified as "`\quad # \hfil`"; therefore the first column will contain left-justified entries, preceded by a `\quad` skip (see §3.5). The remaining four columns follow the pattern "`\hfil #`", so they will be right-justified. Generally, specifying

- `& \hfil # &` generates a right-justified column
- `& # \hfil &` generates a left-justified column
- `& \hfil # \hfil &` generates a centered column
- `& # &` generates an unjustified column (compare with `\rightline`, `\leftline`, `\centerline`, and `\line`; see §3.2)

➡ Each row of ⟨text⟩ must contain the same number of alignment tabs as the ⟨template⟩. Further, the replacement entries in the ⟨text⟩ should correspond to the appearances of # in the ⟨template⟩.

Entries in the template may contain substitution text. This text will appear in all rows of the table.

EXAMPLE 9.7 ▷ Typing

```
{ \bf Saturn History Highlights, 1995-2217 }
\smallskip
\halign{ \hfil # A.D. & \qquad #.\hfil\cr
1997 & Joann di Silva lands on Saturn\cr
2011 & First trading post established near
       the North Pole\cr
2027 & First involuntary colonists are shipped\cr
2112 & “{\it Saturn for Saturnians}”
       party established\cr
2217 & Rent increases force most of
       Saturnians to relocate\cr
}
```

produces

Saturn History Highlights, 1995-2217		
1997	A.D.	Joann di Silva lands on Saturn.
2011	A.D.	First trading post established near the North Pole.
2027	A.D.	First involuntary colonists are shipped.
2112	A.D.	"*Saturn for Saturnians*" party established.
2217	A.D.	Rent increases force most of Saturnians to relocate.

Observe that \cr immediately follows the entries in the second column of ⟨text⟩—otherwise a space would be inserted before the period as it is inserted before "A.D.".

➡ It is best to insert table titles with **\noalign** or **\multispan** commands (see below).

Tables may contain text that is not aligned. Such text is entered with the **\noalign** command.

EXAMPLE 9.8 ▷ Typing

```
{\sc Price and Performance of 1997 car models:}
\halign{ #\hfil & \quad#,000 U.S \$
  &\qquad \hfil #\ m.p.g.\cr
\noalign{\bf Foreign:}
Banana Islands' Jiraffe Royale &32&75\cr
Paraguay's Grandioso &47&72\cr
Taiwan's Uffon &78&69\cr
Japanese Hana &121&68\cr
\noalign{\bf Domestic:}
Fiord's (T)axBreaker &109&71\cr
National Motors' Jap &189&67\cr
General Designs' Jackal &201&82\cr
 }
```

results in

PRICE AND PERFORMANCE OF 1997 CAR MODELS:		
Foreign:		
Banana Islands' Jiraffe Royale	32,000 U.S $	75 m.p.g.
Paraguay's Grandioso	47,000 U.S $	72 m.p.g.
Taiwan's Uffon	78,000 U.S $	69 m.p.g.
Japanese Hana	121,000 U.S $	68 m.p.g.
Domestic:		
Fiord's (T)axBreaker	109,000 U.S $	71 m.p.g.
National Motors' Jap	189,000 U.S $	67 m.p.g.
General Designs' Jackal	201,000 U.S $	82 m.p.g.

The internal variable `\tabskip` controls the spacing between columns. By default, it is set to 0, so that adjacent columns may touch each other. It may be set to any glue value.

EXAMPLE 9.9 ▷ Table

```
\tabskip 3pt
\halign{ \quad #\hfil & \hfil# & \hfil# & \hfil# & \hfil# \cr
 Company & Quarter 1 & Quarter 2 & Quarter 3 & Quarter 4\cr
 Jones \& Sons & 12.32 & 13.34 & 14.32 & 15.11&\cr
 UnderDog Inc  & 33.33 & 35.43 & 36.35 & 45.03&\cr
 \EEFlogo      & 23.33 & 24.34 & 26.12 &  0.72&\cr
 }
```

is typeset as

Company	Quarter 1	Quarter 2	Quarter 3	Quarter 4
Jones & Sons	12.32	13.34	14.32	15.11
UnderDog Inc	33.33	35.43	36.35	45.03
$\mathcal{EEF}$	23.33	24.34	26.12	0.72

while the same table with `\tabskip=6pt` is typeset as

Company	Quarter 1	Quarter 2	Quarter 3	Quarter 4
Jones & Sons	12.32	13.34	14.32	15.11
UnderDog Inc	33.33	35.43	36.35	45.03
$\mathcal{EEF}$	23.33	24.34	26.12	0.72

The tables typeset so far take their natural width. The width can be explicitly specified with the `to` or `spread` parameters of the `\halign` command.

SYNTAX ▷ `\halign to` ⟨dimen⟩ `{`⟨template⟩ `\cr` ⟨text⟩ `\cr` ⟨text⟩ `\cr` ... `\cr }`
▷ `\halign spread` ⟨dimen⟩ `{`⟨template⟩ `\cr` ⟨text⟩ `\cr` ⟨text⟩ `\cr` ... `\cr`
`}`

where `to` makes the table ⟨dimen⟩ wide; `spread` increments or decrements the natural width of the table by ⟨dimen⟩. The `to` and `spread` commands also apply to boxes (see §8.3).

The tables are expanded by stretching the `\tabskip` glue. Therefore, specifying `\halign to` or `\halign spread` will have an effect only if the `\tabskip` glue is stretchable, i.e., has a `plus` component (see §5.4).

EXAMPLE 9.10 ▷ In the previous example, inserting the line

```
\tabskip  5pt plus 1fill
```

before the `\halign` command and replacing `\halign{` with

```
\halign to 320pt{
```

results in

Company	Quarter 1	Quarter 2	Quarter 3	Quarter 4
Jones & Sons	12.32	13.34	14.32	15.11
UnderDog Inc	33.33	35.43	36.35	45.03
$\mathcal{EEF}$	23.33	24.34	26.12	0.72

while the same table with specification `\halign to 305pt` appears as

Company	Quarter 1	Quarter 2	Quarter 3	Quarter 4
Jones & Sons	12.32	13.34	14.32	15.11
UnderDog Inc	33.33	35.43	36.35	45.03
$\mathcal{EEF}$	23.33	24.34	26.12	0.72

➡ The `\tabskip` glue applies to all columns of the table. To use different values of `\tabskip` for different columns, `\tabskip` may be set and reset inside the ⟨template⟩.

9.4 Using Rules In Tables

TOPICS ▷ `\offinterlineskip`, `\strut`, `\multispan`, `\omit`

The appearance of a table is greatly enhanced by the presence of rules. The easiest way to typeset vertical rules is to specify them as part of the ⟨template⟩ and reserve a separate column for each rule. The `\offinterlineskip` and `\strut` commands are used in the next example; their actions are described below.

EXAMPLE 9.11 ▷ Typing

```
{\offinterlineskip
\tabskip 3pt plus 1fill
\halign to 379pt{
   \strut#
   &\vrule#& \quad # \hfil
   &\vrule#& \hfil #
   &\vrule#& \hfil #
```

```
   &\vrule#&& \hfil #
   &\vrule#&& \hfil #
   &\vrule# \cr
& Company && Quarter 1 && Quarter 2
          && Quarter 3 && Quarter 4 &\cr
& Jones \& Sons && 12.32 && 13.34 && 4.32 && 15.11 &\cr
& UnderDog Inc  && 33.33 && 35.43 && 6.35 && 45.03 &\cr
& \EEFlogo      && 23.33 && 24.34 && 6.12 &&  0.72 &\cr
  } }
```

generates the table

Company	Quarter 1	Quarter 2	Quarter 3	Quarter 4
Jones & Sons	12.32	13.34	14.32	15.11
UnderDog Inc	33.33	35.43	36.35	45.03
$\mathcal{EEF}$	23.33	24.34	26.12	0.72

The empty entries in the ⟨text⟩ of the table correspond to the `\vrule` specifications in the ⟨template⟩.

In this example, `\strut` is used to make all lines `8.5pt` high—this guarantees uniformity of the table. `\offinterlineskip` suppresses the extra vertical spacing that TEX normally inserts between lines. Without the `\strut` command the table becomes

Company	Quarter 1	Quarter 2	Quarter 3	Quarter 4
Jones & Sons	12.32	13.34	14.32	15.11
UnderDog Inc	33.33	35.43	36.35	45.03
$\mathcal{EEF}$	23.33	24.34	26.12	0.72

and without both `\strut` and `\offinterlineskip` it changes into

Company	Quarter 1	Quarter 2	Quarter 3	Quarter 4
Jones & Sons	12.32	13.34	14.32	15.11
UnderDog Inc	33.33	35.43	36.35	45.03
$\mathcal{EEF}$	23.33	24.34	26.12	0.72

In this example, the interline glue breaks the vertical rules; in the preceding example, the absence of `\strut` causes the rows to collide.

➡ Observe that `\offinterlineskip` has to be specified only once, while `\strut`, because it is listed in the ⟨template⟩, is specified for each line.

Horizontal rules can be inserted using the `\noalign` command. To ensure that the length of these rules corresponds to the length of the table, the entire table should be embedded in a vertical box.

EXAMPLE 9.12 ▷ Typing

```
\vbox{
\offinterlineskip
\tabskip 3pt plus 1fill
\halign{
   \strut#
   &\vrule#& \quad # \hfil
   &\vrule#& \hfil #
   &\vrule#& \hfil #
   &\vrule#& \hfil #
   &\vrule#& \hfil #
   &\vrule# \cr
   \noalign{\hrule}  % top rule
& Company && Quarter 1 && Quarter 2
         && Quarter 3 && Quarter 4 &\cr
   \noalign{\hrule}
& Jones \& Sons && 12.32 && 13.34 && 4.32 && 15.11 &\cr
   \noalign{\hrule}
& UnderDog Inc  && 33.33 && 35.43 && 6.35 && 45.03 &\cr
   \noalign{\hrule}
& \EEFlogo      && 23.33 && 24.34 && 6.12 &&  0.72 &\cr
   \noalign{\hrule} % bottom rule
  }}
```

results in

Company	Quarter 1	Quarter 2	Quarter 3	Quarter 4
Jones & Sons	12.32	13.34	14.32	15.11
UnderDog Inc	33.33	35.43	36.35	45.03
$\mathcal{EEF}$	23.33	24.34	26.12	0.72

Entries that span alignment positions can be inserted with the `\multispan` command.

SYNTAX ▷ `\multispan` ⟨number⟩

This command specifies that the following entry should be stretched over the next ⟨number⟩ columns. ⟨number⟩ should be a positive integer number.

EXAMPLE 9.13 ▷ In the previous example, inserting

```
& \multispan9 \hfil
 Market Shares of Food Surrogate Manufacturers \hfil&\cr
\noalign{\hrule}
```

after the line marked '`%top rule`' changes the table into

Market Shares of Food Surrogate Manufacturers				
Company	Quarter 1	Quarter 2	Quarter 3	Quarter 4
Jones & Sons	12.32	13.34	14.32	15.11
UnderDog Inc	33.33	35.43	36.35	45.03
$\mathcal{EEF}$	23.33	24.34	26.12	0.72

The `\multispan` entry in the above example spans 9 columns: 5 text columns and 4 `\vrule` columns.

➡ Use `\multispan{11}` and not `\multispan11` to generate an entry spanning eleven columns. The second specification would result in an entry spanning 1 column; the extra digit '1' would be appended to the beginning of the entry.

EXAMPLE 9.14 ▹ Thicker rules or double rules may be used to enhance the appearance of a table.

```
\offinterlineskip
\tabskip 0pt
\halign{
   \strut
   \vrule width 2pt #& \quad # \hfil
   &\vrule#& \hfil #
   &\vrule#& \hfil #
   &\vrule#& \hfil #
   &\vrule#& \hfil #
   &\vrule width 2pt# \cr
\noalign{\hrule height 2pt}  % top rule
& \multispan9
\hfil Market Shares of Food Surrogate Manufacturers
  \hfil&\cr
\noalign{\hrule}  % top rule
& Company&&\ Quarter 1&&\ Quarter 2&&\
             Quarter 3&&\ Quarter 4 &\cr
\noalign{\hrule}
& Jones \& Sons && 12.32 && 13.34 && 14.32 && 15.11 &\cr
\noalign{\hrule}
& UnderDog Inc  && 33.33 && 35.43 && 36.35 && 45.03 &\cr
\noalign{\hrule}
& \EEFlogo      && 23.33 && 24.34 && 26.12 &&  0.72 &\cr
\noalign{\hrule height 2pt}  % bottom rule
  }
```

produces

Market Shares of Food Surrogate Manufacturers				
Company	Quarter 1	Quarter 2	Quarter 3	Quarter 4
Jones & Sons	12.32	13.34	14.32	15.11
UnderDog Inc	33.33	35.43	36.35	45.03
$\mathcal{EEF}$	23.33	24.34	26.12	0.72

The `\hrule height 2pt` and `\vrule width 2pt` commands specify thicker rules. The default rule width is 0.4pt.

Chapter

10

Font Rotation

Of all the things that can be done with TEX, font rotation is perhaps one of the most fascinating. The original idea is due to Alan Hoenig, who was kind enough to allow an abridged version of his article to appear as the next section of this book. Under standard TEX, font rotation requires massive use of METAFONT for preparation of rotated fonts. Under VTEX, as the reader will see in a subsequent section, the task is largely handled by VTEX and drivers.

10.1 Circular Typesetting with TEX and METAFONT–by Alan Hoenig

METAFONT is a companion program to TEX. The author of TEX used it to design and create the Computer Modern font family that accompanies many implementations of TEX. Although METAFONT is a sophisticated tool just right for creating digital fonts, it can also be used with great profit to create graphic images for inclusion within TEX documents. Like TEX, METAFONT is also great fun to use.

The purpose of this brief chapter is to show how it is possible to use the computer modern font files together with a dash of generality to create fonts which make it possible to typeset on circles. A more detailed treatment of this subject appeared in the article *Circular Reasoning,* in the journal *TUGboat,* **11** (2), pp. 183-190 (June, 1990).

What does it mean to place letters on a circle? Here's my interpretation. I imagine dividing the circle into many, many wedges of equal sizes. Furthermore, I imagine lines connecting the two "corners" of each wedge (the places where the wedge intersects the circle). Let's call each of these imaginary lines a *face*. (Mathematicians know them as *chords* of the circle.) Circular typesetting means centering a single letter on each face.

TEX can't actually lay letterforms on their sides. What we do therefore is demand that METAFONT create a family of new fonts, each member of a new font at the proper orientation for one particular face of our circle. Once that is done, it's a straightforward matter to design TEX macros that select individual letters from a message and position them properly on the circle.

In this chapter, you will see how to create the family of fonts and how to use the proper TEX macros. Bear in mind that circular typesetting is a compromise over TEX's normally impeccable typesetting job. Since each letter is centered on the face, narrow letters like 'I' and '1' will have more space on either side of them than will wide letters like 'M' and 'W'. Furthermore, all ligatures are broken, and there is no kerning between adjacent letters.

For that reason, you may wish to use the TEX `typewriter font` as the font to rotate. In this font, all characters normally occupy the same widths. In what follows I use `cmtt12`, the twelve-point typewriter type, to generate rotated fonts for this reason.

The recipes in this chapter assume that you want to typeset **bold extended** type along the circle. If all faces on the circle are exactly 12 pt wide, then we can show that the radius of the circle must be 61.21 pt. (This makes the diameter of the circle a shade under 1.7 in.)

- METAFONT Font Files

It's helpful to begin by describing the three-tiered structure of METAFONT font files. We create a font by typing a command something like

```
metafont cmr10
```

which is directly analogous to a TEX typesetting command like **`tex mydoc`**. Here, `cmr10.mf` is a *parameter file*. It contains values for the parameters which METAFONT uses to construct individual fonts in the Computer Modern family. It turns out that there are about 60 parameters we can adjust. Giving different values to these parameters generates all the fonts in the standard suite of fonts that accompany any healthy TEX distribution.

The parameter file will then call a *driver file* into play. It is the purpose of this short file to organize the font into uppercase and lowercase portions, portions containing the special characters, and so on.

As part of its job, the driver file calls upon one or more *program files*, which contain the actual instructions for creating the letterforms in each font. These instructions are given using the grammar and syntax of the METAFONT programming language, and are given in terms of the sixty or so parameters whose values are fixed in the parameter file.

The output of any METAFONT job is a generic pixel file. If you create fonts suitable for a laser printer, part of the instructions you give in the initial command line must tell METAFONT to make fonts with a value of

```
mode=hplaser
```

or something like that. `mode` is a METAFONT command to set the particular values of certain parameters that are appropriate for your particular printer. Here, I imagine that the value **`hplaser`** sets these values for the Hewlett-Packard LaserJet printer, a popular 300 dpi laser printer.

When this job finishes, I will find a font file called `cmr10.300` and a font metric file `cmr10.tfm` on my computer. This pixel file is not quite suitable for use. First, we convert it to a *packed pixel file* which uses much less space on your hard disk. Do this via the command

```
gftopk cmr10.300 cmr10.pk
```

and then delete `cmr10.300`. (The program `gftopk` should be part of your METAFONT distribution.) Now put the **`tfm`** file with all the other tt tfm files and put the **`pk`** file with all the other pixel files suitable for printing at an unmagnified output. On a PC-type machine, this will be in a directory with a name something like `\tex\pixels\dpi300` or like `\tex\pixels\1500`.

- Making Changes to the METAFONT Font Files

The rotated printing looks best when we use a monospaced font like `cmtt12`, so we will describe the process for generating rotated copies of these fonts.

Begin by creating 32 copies of the parameter file `cmtt12.mf`, naming them `cmtt1200.mf` through `cmtt1231.mf`. Also make a copy of the driver file `roman.mf` and call it `roroman.mf`. (It will be a *ro*tated Roman driver.)

At the beginning of each parameter file, change the comments to remind yourself of the changes you will have made to it. At the end of the file, change the statement '`generate roman;`' to

```
generate roroman;
```

Immediately preceding this line, enter these lines very carefully. Be careful to distinguish carefully between semi-colons and full colons in particular!

```
numeric wedge_angle; wedge_angle=360/32;
numeric face; face=0;
numeric rotation_angle;
  rotation_angle=90-(face+.5)*wedge_angle;
fontdimen9: face, rotation_angle;
fontdimen11: sind wedge_angle, cosd wedge_angle;
fontdimen13: sind(rotation_angle), cosd(rotation_angle);
 % for r=62.21pt (the largest circle)
fontdimen15: sind(90-2(face+.5)*wedge_angle),
 cosd(90-2(face+.5)*wedge_angle); % for r=30.61pt
fontdimen17: sind(90-4(face+.5)*wedge_angle),
 cosd(90-4(face+.5)*wedge_angle); % for r=15.30pt
```

Actually, *make sure the second line is varied in each of your parameter files.* In `cmtt1200.mf`, set `face=0;`. In file `cmtt1201.mf`, set `face=1;`, and so on, until `face=31;` in file `cmtt1231.mf`.

- Changes to `roroman.mf`

Now go to file `roroman.mf`. Look for the line

```
mode_setup; font_setup;
```

and immediately before this line insert

```
currenttransform:=currenttransform rotated rotation_angle;
def t_=transformed currenttransform enddef;
```

Check your spelling carefully, and then save the file. Now you're ready to create the new fonts.

- Thirty-Two New Fonts

To generate the fonts, you have to run METAFONT 32 times! Each time, the command line will be something like

```
mf &cm \mode=hplj; input cmtt1211
```

Here, the `mode` command makes sure that the parameters appropriate for your printer control the font. (In this example, `hplj` stands for Hewlett Packard LaserJet, but your version of METAFONT will have different `mode` values.) Of course, the file name following the word `input` should vary from `cmtt1200` through `cmtt1231`.

You may wish to generate only the uppercase characters in each font, and thereby save time and disk space. To do that, refer again to the contents of file `roroman.mf`, and comment out with a percent sign % all but the first of the `input` lines in that file.

The result of the METAFONT process will be a series of pixel files with names like `cmtt1200.300` and font metric files like `cmtt00.tfm`. Each of them has to be converted to a packed pixel file via commands of the form

```
gftopk cmtt1200
```

which generates files with names like `cmtt00.pk`. Now move the `pk` files to the proper subdirectories on your hard disk. These directories will have a name similar to

```
\tex\pixel\dpi300
```

but carefully check the documentation that accompanies your implementation of TeX. Also, move all the `tfm` files to a directory called

```
\tex\tfm
```

or something similar. Finally, in the interests of good housekeeping, delete the `log` files and other detritus in the place where you did your METAFONTing.

Typesetting with these fonts requires a series of special-purpose macros. Here they are. When you enter these macros in your document file, make sure you have copied them *exactly* as they appear!

```
{\catcode`p=12 \catcode`t=12 \gdef\\#1pt{#1}}
 \let\getfactor=\\

\def\dolist{\afterassignment\dodolist \let\next= }
\def\dodolist{\ifx\next\endlist \let\next=\relax
 \else \\\let\next=\dolist \fi
```

```
 \next}
\def\endlist{\endlist }

\def\\{\expandafter\if\space\next \newcoords
   \else\point\next \fi}

\newcount\faceno \newdimen\Bigradius \newdimen\Radius
\newdimen\radius \newdimen\r
\Bigradius=61.21pt \Radius=.5\Bigradius \radius=.5\Radius
\r=\Bigradius
\newdimen\x \newdimen\y
\def\initialize{\font\anglefont=cmtt1200 \global\faceno=0
 \x=-\r \y=0pt }
\def\nextpointt{%
 \dimen0=\expandafter\getfactor \the\fontdimen12\anglefont \x
 \dimen2=\expandafter\getfactor \the\fontdimen11\anglefont \y
 \advance\dimen0 by\dimen2 \dimen1=\dimen0
 \dimen0=-\expandafter\getfactor \the\fontdimen11\anglefont \x
 \dimen2= \expandafter\getfactor \the\fontdimen12\anglefont \y
 \advance\dimen0 by\dimen2 \global\x=\dimen1\global\y=\dimen0}
\def\nextpoint{\nextpointt \prepareforenextface}
\let\newcoords=\nextpoint

\newcount\lastface \lastface=31
\def\prepareforenextface{\global\advance\faceno by1
 \ifnum\faceno>\lastface \global\faceno=0
 \message{Too many letters in your circular message!}%
  \else \ifnum\faceno<9 \font\anglefont=cmtt120\the\faceno
    \else \font\anglefont=cmtt12\the\faceno \fi
 \fi}

\newdimen\xprime \newdimen\yprime
\def\setonbase{\xprime=\x \yprime=\y \baseoffset=.5\base
 \advance\baseoffset by-.5\wd2 %current char in box2
 \dimen0=\expandafter\getfactor\the\fontdimen14\anglefont
   \baseoffset
 \dimen2=\expandafter\getfactor\the\fontdimen13\anglefont
\baseoffset
 \advance\xprime by \dimen0 \advance\yprime by\dimen2 }

\newdimen\base \base=12pt \newdimen\baseoffset
\newtoks\currchar
\def\point#1{\currchar=\expandafter{#1}%
 \setbox2=\hbox{\anglefont \the\currchar}\setonbase
 \wd2=0pt \ht2=0pt \dp2=0pt
```

```
\rlap{\kern\xprime \raise\yprime\box2 }\newcoords}

\def\circumtype#1{\initialize
 \setbox0=\hbox{\dolist#1\endlist}\leavevmode \box0 }
```

It's easy to use these falsely intimidating macros. You may set type on a circle by entering a command of the form

```
\circumtype{ABCDEFGHIJKLMNOPQRSTUVWXYZABCDEF}
```

Have fun!

10.2 Circular Typesetting with VTEX

The necessity of pre-creation of a multitude of rotated fonts with conventional TEX is the direct cause of the use of bitmap fonts. VTEX, which uses vector fonts can rotate fonts on the fly. The macros presented is this section are adaptations of those presented above and in the original Hoenig article. Before proceeding with macro listings, we will summarize the process.

First, to be able to position rotated characters, TEX needs to know the sine's and cosine's of the rotation angles. Generally, TEX has no built-in facility for computing these values. While it is entirely possible to compute a power series using TEX's arithmetic capabilities, this technique will be unacceptably slow. In the previous section, Hoenig used the numeric capabilities of METAFONT to compute the values and then passed them to TEX via font dimensions. VTEX, instead, was expanded to include an additional primitive called `\sincos`. `\sincos` takes an argument representing an angle and returns its sine and cosine in the `\sine` and `\cosine` registers. For example, typing

```
\sincos90pt
```

sets `\sine` to 1pt and `\cosine` to 0pt. Observe that `\sine`, `\cosine`, and the argument to `\sincos` are all dimensions with the convention that 1° means 1pt.

The computed `\sine` and `\cosine` would also be needed by drivers to actually rotate the fonts. VTEX transmits these values to the drivers via a `\special{R...}` command. To specify the angle of rotation, we type

```
\special{R\the\cosine,\the\sine}
```

while to turn it off, we type

```
\special{R0,0}
```

With these tools at hand, we shall now proceed to plagiarize Hoenig's work:

```
{\catcode`p=12 \catcode`t=12 \gdef\\#1pt{#1}}
```

```
\let\getfactor=\\

\newdimen\x \newdimen\cos
\newdimen\y \newdimen\sin
\def\initialize{\global\x=0pt\global\y=0pt}

\def\action{}
\def\dolist{\afterassignment\dodolist\let\next= }
\def\dodolist{\ifx\next\endlist
  \let\next\relax
  \else\action
  \\\let\next\dolist\fi
  \next}
\def\endlist{\endlist}

\def\\{\expandafter\if\space\next\addspace
          \else\point\next\fi}

\newbox\spacebox\setbox\spacebox=\hbox{\ }
 \def\addspace{\setbox0=\copy\spacebox\newcoords}

\def\point#1{%
  \setbox0=\hbox{#1}% used by \newcoords
  \setbox2=\hbox{#1}% used for typesetting
   \wd2=0pt \ht2=0pt \dp2=0pt
   \rlap{\kern\x \raise\y \box2}%
   \newcoords}
 \def\newcoords{%
   \global\advance\x by  \cos
   \global\advance\y by -\sin}

 \def\angletype#1{\initialize
   \leavevmode\setbox1=
     \hbox{\dolist#1\endlist}\box1}
```

Now we can safely type

```
\def\tryrotation#1{%
\setrotation #1pt%
\def\sin{\expandafter\getfactor\the\sine\wd0}%
\def\cos{\expandafter\getfactor\the\cosine\wd0}%
\special{R\the\cosine,\the\sine}%
\angletype{This text is rotated #1 degrees}%
\special{R0,0}}

\vskip-1cm
```

```
\hskip7cm
\font\anglefont=mvhand \anglefont
\tryrotation{60}%
\tryrotation{150}%   Spaces are removed to keep
\tryrotation{240}%   the reference point
\tryrotation{330}%   the same for all four lines.
```

obtaining

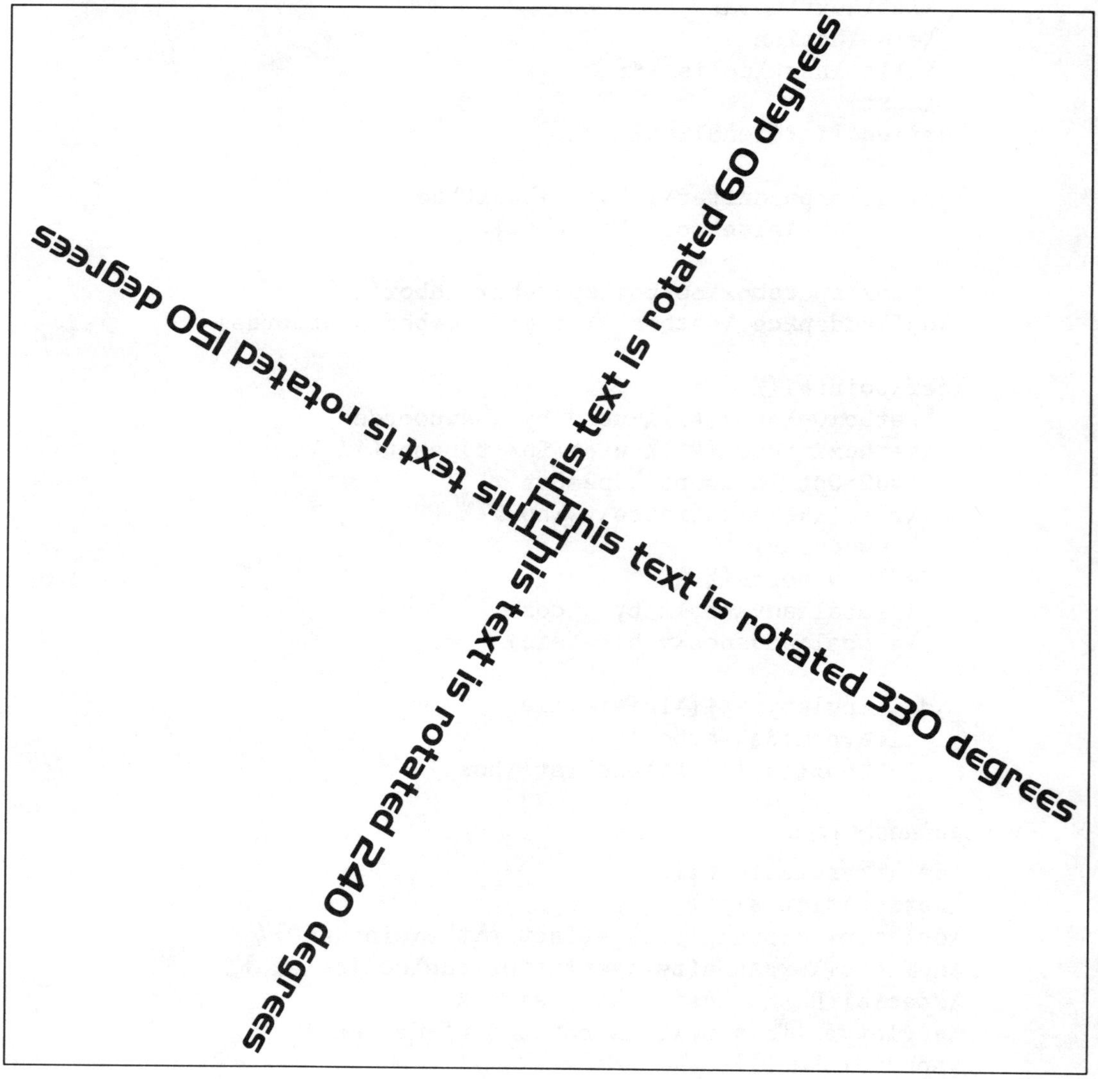

The next example finally illustrates circular typesetting. The `\circletype` macro we will shortly present will typeset text on a circle. The radius of the circle is never computed and is determined solely by the length of the text. Three global variables we will need are the number of characters in the string (`\charcount`), current angle of rotation (`\rotvalue`), and the increment for the rotation angle to apply after each character (`\rotdelta`). The general appearance of `\circletype` is

```
\newcount\charcount
\newdimen\rotdelta
\newdimen\rotvalue

\def\circletype#1{\anglefont
  \setbox0=\hbox{\countchars{#1}}%
  \rotdelta=-360pt% i.e. degrees
  \divide\rotdelta by \charcount
  \rotvalue=0pt
  \setrotation \rotvalue
  \def\sin{\expandafter\getfactor\the\sine\wd0}%
  \def\cos{\expandafter\getfactor\the\cosine\wd0}%
  \special{R\the\cosine,\the\sine}%
  \angletype{#1}%
  \special{R0,0}}
```

where `\countchars` computes the number of characters and stores them in `\charcount`. We reuse the `\dolist` idea with a new `\action`:

```
\def\countchars#1{{%
\global\charcount=0
  \def\action{\global\advance\charcount by 1}
   \dolist#1\endlist
   }}
```

Finally, typing

```
\font\anglefont=mvtt10 scaled 2400
\hskip4cm \circletype{The circles are forever.\ }
```

produces

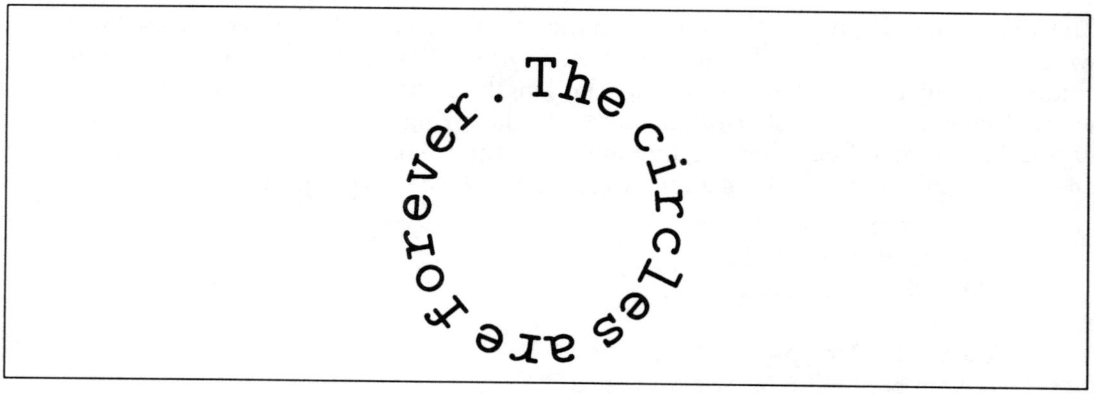

Many modification of these macros are possible; the simplest is the replacement of `\rotdelta=-360pt` by `\rotdelta=360pt` which will lead to letters facing inward:

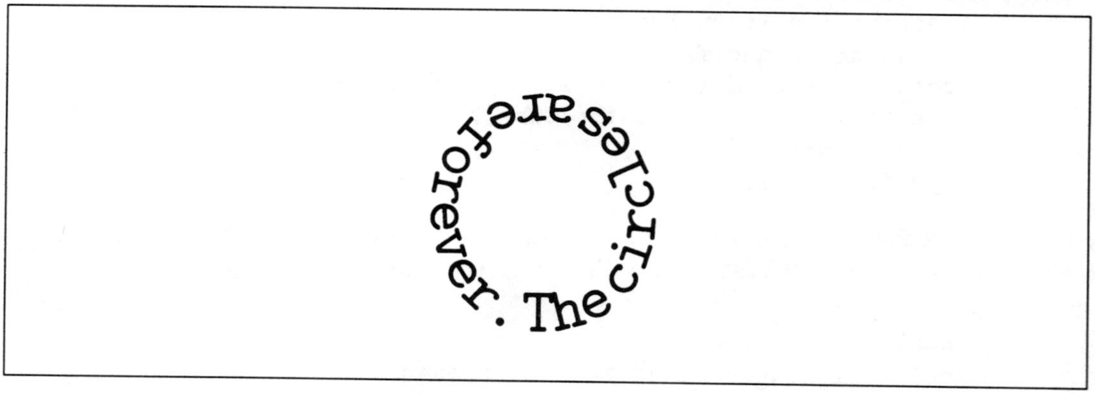

➡ The circles above, although appear to be of different size, actually have the same "baseline radius."

It is also possible to alter (and improve) the interword spacing (the last example will definitely benefit), typeset text on an ellipse, a spiral, or any other curve. One can further change the size of the type as it moves along the curve and so on. And, as first observed by Hoenig, it is also possible to have fun doing all this. Indeed.

Chapter 11

Appendix

11.1 Tables of Mathematical Symbols

There are eight categories of mathematical symbols (see §4.18). Special symbols occur only in the following six categories: **ORD** (Ordinary Symbols), **OP** (Large Operators), **BIN** (Binary Operators), **REL** (Relational Operators), and **OPEN** and **CLOSE** (Opening and Closing Punctuation Symbols).

Category **ORD**: Ordinary Symbols

All letters and digits fall into this category. The following additional symbols are available:

Greek Letters (Category **ORD**)					
Uppercase		Lowercase		Alternate Lowercase	
Symbol	Command	Symbol	Command	Symbol	Command
A	[1]	α	\alpha		
B	[1]	β	\beta		
Γ	\Gamma	γ	\gamma		
Δ	\Delta	δ	\delta		
E	[1]	ϵ	\epsilon	ε	\varepsilon
Z	[1]	ζ	\zeta		
H	[1]	η	\eta		
Θ	\Theta	θ	\theta	ϑ	\vartheta
I	[1]	ι	\iota		
K	[1]	κ	\kappa		
Λ	\Lambda	λ	\lambda		
M	[1]	μ	\mu		
N	[1]	ν	\nu		
Ξ	\Xi	ξ	\xi		
O	[1]	o	[2]		
Π	\Pi	π	\pi	ϖ	\varpi
P	[1]	ρ	\rho	ϱ	\varrho
Σ	\Sigma	σ	\sigma	ς	\varsigma
T	[1]	τ	\tau		
Υ	\Upsilon	υ	\upsilon		
Φ	\Phi	ϕ	\phi		
X	[1]	χ	\chi		
Ψ	\Psi	ψ	\psi		
Ω	\Omega	ω	\omega		

1) Identical to the corresponding capital letter in Roman Font.
2) Lowercase *omicron* is identical to lowercase *o* in Math Italics.

➡ To obtain slanted capital Greek letters, use the prefix `\mit` (`\mit\Gamma` yields $\mit\Gamma$). The `\mit` prefix may also be used to produce old-style digits: `$\mit0123456789$` typesets as $\mit0123456789$. `\mit` may also be used in text mode.

The capital calligraphic letters $\mathcal{A}$, $\mathcal{B}$, $\mathcal{C}$, $\mathcal{D}$, $\mathcal{E}$, $\mathcal{F}$, $\mathcal{G}$, $\mathcal{H}$, $\mathcal{I}$, $\mathcal{J}$, $\mathcal{K}$, $\mathcal{L}$, $\mathcal{M}$, $\mathcal{N}$, $\mathcal{O}$, $\mathcal{P}$, $\mathcal{Q}$, $\mathcal{R}$, $\mathcal{S}$, $\mathcal{T}$, $\mathcal{U}$, $\mathcal{V}$, $\mathcal{W}$, $\mathcal{X}$, $\mathcal{Y}$, and $\mathcal{Z}$ are obtained by using the `\cal` prefix in math mode (for example, `$\cal A$` produces $\mathcal{A}$).

CAUTION ▷ Use `\cal` only with capital letters.

More math symbols (Category **ORD**)					
Symbol	Command	Symbol	Command	Symbol	Command
∂	`\partial`	$\hbar$	`\hbar`	$\Im$	`\Im`
$\imath$	`\imath`	$\jmath$	`\jmath`	$\aleph$	`\aleph`
$\wp$	`\wp`	ℓ	`\ell`	$\Re$	`\Re`
$\flat$	`\flat`	$\natural$	`\natural`	$\sharp$	`\sharp`
$\forall$	`\forall`	$\exists$	`\exists`	∞	`\infty`
$\triangle$	`\triangle`	∇	`\nabla`	$\emptyset$	`\emptyset`
$\bot$	`\bot`	$\top$	`\top`	$\backslash$	`\backslash`
$\neg$	`\neg`	$\angle$	`\angle`	$\surd$	`\surd`

The dotless letters $\imath$ (`\imath`) and $\jmath$ (`\jmath`) are used when i and j are accented.

EXAMPLE 11.1 ▷ `$\check\jmath$` produces $\check\jmath$.

Card suits (Category **ORD**)			
Symbol	Command	Symbol	Command
$\spadesuit$	`\spadesuit`	$\heartsuit$	`\heartsuit`
$\diamondsuit$	`\diamondsuit`	$\clubsuit$	`\clubsuit`

Category **OP**: Large Math Operators

The symbols in this category come in two sizes: small for ordinary math and large for display math.

Large Operators (Category **OP**)					
Symbol	Command	Symbol	Command	Symbol	Command
$\coprod$, $\coprod$	\coprod	$\bigcup$, $\bigcup$	\bigcup	$\bigoplus$, $\bigoplus$	\bigoplus
$\bigvee$, $\bigvee$	\bigvee	$\int$, $\int$	\intop	$\bigodot$, $\bigodot$	\bigodot
$\bigwedge$, $\bigwedge$	\bigwedge	$\prod$, $\prod$	\prod	$\oint$, $\oint$	\ointop
$\biguplus$, $\biguplus$	\biguplus	$\sum$, $\sum$	\sum	$\bigsqcup$, $\bigsqcup$	\bigsqcup
$\bigcap$, $\bigcap$	\bigcap	$\bigotimes$, $\bigotimes$	\bigotimes		

These symbols can receive sub- or superscripts, whose placement is controlled by the commands `\limits` and `\nolimits` (see §4.4). `\int` and `\oint` are alternate names for `\intop` and `\ointop`. `\int` and `\oint` use the `\nolimits` form of the scripts, both in ordinary math and in display math. `\intop` and `\ointop`, as well as the other math operators, use `\nolimits` scripts in ordinary math and `\limits` scripts in display math.

Category **BIN**: Binary Operators

Binary Operators (Category **BIN**)					
Symbol	Command	Symbol	Command	Symbol	Command
$+, -$	+,-	$\cap$	\cap	$\wedge$	\wedge
$\pm$	\pm	$\cup$	\cup	$\oplus$	\oplus
$\mp$	\mp	$\uplus$	\uplus	$\ominus$	\ominus
$\setminus$	\setminus	$\sqcap$	\sqcap	$\otimes$	\otimes
$\cdot$	\cdot	$\sqcup$	\sqcup	$\oslash$	\oslash
$\times$	\times	$\triangleleft$	\triangleleft	$\odot$	\odot
$\ast$	\ast	$\triangleright$	\triangleright	$\dagger$	\dagger
$\star$	\star	$\wr$	\wr	$\ddagger$	\ddagger
$\diamond$	\diamond	$\bigcirc$	\bigcirc	$\amalg$	\amalg
$\circ$	\circ	$\bigtriangleup$	\bigtriangleup	$\otimes$	\vee
$\bullet$	\bullet	$\bigtriangledown$	\bigtriangledown	$\div$	\div

`\dagger` and `\ddagger` are also available in text mode as `\dag` and `\ddag`. `\wedge` and `\vee` are also known as `\land` (logical and) and `\lor` (logical or). VTEX fonts also include daggers in all standard fonts; these are accessed in the text mode with the ligatures `+!` and `+!!`.

Category **REL**: Relations

Relations (Category **REL**)					
Symbol	Command	Symbol	Command	Symbol	Command
$<$	`<`	$\subset$	`\subset`	$\equiv$	`\equiv`
$>$	`>`	$\supset$	`\supset`	$\simeq$	`\simeq`
$=$	`=`	$\subseteq$	`\subseteq`	$\approx$	`\approx`
$\leq$	`\leq`, `\le`	$\supseteq$	`\supseteq`	$\asymp$	`\asymp`
$\geq$	`\geq`, `\ge`	$\sqsubseteq$	`\sqsubseteq`	$\cong$	`\cong`
$\ll$	`\ll`	$\sqsupseteq$	`\sqsupseteq`	$\propto$	`\propto`
$\gg$	`\gg`	$\vdash$	`\vdash`	$\bowtie$	`\bowtie`
$\prec$	`\prec`	$\dashv$	`\dashv`	$\doteq$	`\doteq`
$\preceq$	`\preceq`	$\in$	`\in`	$\models$	`\models`
$\succ$	`\succ`	$\ni$	`\ni`, `\owns`	$\perp$	`\perp`
$\succeq$	`\succeq`	$\mid$	`\mid`, `\vert`	$\smile$	`\smile`
$\sim$	`\sim`	$\parallel$	`\parallel`	$\frown$	`\frown`

The `\not` prefix can be used to generate negated relations. For example, typing `\not=` produces $\boxed{\neq}$, and `\not\subset` produces $\boxed{\not\subset}$. `\not=` can also be abbreviated as `\ne` or `\neq`. Negated relations are obtained by superimposing a slash over a relation symbol; in some cases the results may not be perfect. In particular, `\not\in` typesets as $\boxed{\not\in}$; use instead `\notin`, which typesets as $\boxed{\notin}$. Using the `\kern` command you can design variations of the $\boxed{\ll}$ symbol with different distances between the less signs. For instance, typing `$<\kern-3pt<$` produces $\boxed{<\kern-3pt<}$.

Arrows (Category **REL**)			
Syml	Command	Syml	Command
$\leftarrow$	\leftarrow, \gets	$\uparrow$	\uparrow[1]
$\rightarrow$	\rightarrow, \to	$\downarrow$	\downarrow[1]
$\leftrightarrow$	\leftrightarrow	$\Uparrow$	\Uparrow[1]
$\Leftarrow$	\Leftarrow	$\Downarrow$	\Downarrow[1]
$\Rightarrow$	\Rightarrow	$\updownarrow$	\updownarrow[1]
$\Leftrightarrow$	\Leftrightarrow	$\Updownarrow$	\Updownarrow[1]
$\mapsto$	\mapsto	$\nearrow$	\nearrow
$\hookleftarrow$	\hookleftarrow	$\searrow$	\searrow
$\leftharpoonup$	\leftharpoonup	$\swarrow$	\swarrow
$\leftharpoondown$	\leftharpoondown	$\nwarrow$	\nwarrow
$\longleftarrow$	\longleftarrow	$\Longleftrightarrow$	\Longleftrightarrow
$\longrightarrow$	\longrightarrow	$\longmapsto$	\longmapsto
$\longleftrightarrow$	\longleftrightarrow	$\hookrightarrow$	\hookrightarrow
$\Longleftarrow$	\Longleftarrow	$\rightharpoonup$	\rightharpoonup
$\Longrightarrow$	\Longrightarrow	$\rightharpoondown$	\rightharpoondown
$\rightleftharpoons$	\rightleftharpoons		

1) These arrows can be stretched (see §4.11).

Notice that capitalization doubles the shaft of the arrow.

\iff is a synonym for \Longleftrightarrow, except that it also adds extra space at each end of the arrow.

Categories **OPEN** & **CLOSE**: Delimiters

Categories **OPEN & CLOSE**			
Opening		Closing	
Symbol	Command	Symbol	Command
$[$	\lbrack	$]$	\rbrack
$\lfloor$	\lfloor	$\rfloor$	\rfloor
$\lceil$	\lceil	$\rceil$	\rceil
$\{$	\lbrace, \{	$\}$	\rbrace, \}
$\langle$	\langle	$\rangle$	\rangle

Additional details can be found in §4.8 (Delimiters).

11.2 Text Font Layouts

While TEX currently supports fonts that contain up to 256 characters, CM fonts used with standard TEX have only 128 or fewer characters apiece. This is because they date back to the days when TEX itself could not handle fonts with more than 128 characters. The 128 character restriction led to omission of many important characters from the CM fonts; it is also the cause of different layouts in different CM fonts. VTEX fonts, even those that are based on CM prototypes, all follow the same layout.

In most of the fonts the positions of normally unprintable characters, such as control characters or spaces, are used for Greek letters, ligatures, and accents.

The teletype fonts (`cmtt`, `cmsltt`, `cmitt`, `cmvtt`, and `cmtcsc`) have the layout closest to the 7-bit ASCII. Here is the layout of the typewriter font `cmtt10`:

cmtt10

	0	1	2	3	4	5	6	7	8	9
	Γ	Δ	Θ	Λ	Ξ	Π	Σ	Υ	Φ	Ψ
1x	Ω	↑	↓	'	ﬃ	ﬄ	ı	ȷ	`	´
2x	ˇ	˘	¯	˚	¸	ß	æ	œ	ø	Æ
3x	Œ	Ø	␣	!	"	#	$	%	&	'
4x	(	)	*	+	,	-	.	/	0	1
5x	2	3	4	5	6	7	8	9	:	;
6x	<	=	>	?	@	A	B	C	D	E
7x	F	G	H	I	J	K	L	M	N	O
8x	P	Q	R	S	T	U	V	W	X	Y
9x	Z	[	\	]	^	_	‘	a	b	c
10x	d	e	f	g	h	i	j	k	l	m
11x	n	o	p	q	r	s	t	u	v	w
12x	x	y	z	{	|	}	~	¨		

Notice the layout lacks the text ligatures (unneeded in unproportional font), double opening and closing quotes, and Polish Ł, normally in position 32.

Most of the other text fonts lack a few punctuation characters that do not normally appear in text. These characters include the backslash \, braces { and }, and the double quote " characters. Here is the layout of the ordinary roman font `cmr10`:

`cmr10`

	0	1	2	3	4	5	6	7	8	9
	Γ	Δ	Θ	Λ	Ξ	Π	Σ	Υ	Φ	Ψ
1x	Ω	ff	fi	fl	ffi	ffl	ı	ȷ	`	´
2x	ˇ	˘	¯	˚	¸	ß	æ	œ	ø	Æ
3x	Œ	Ø	␣	!	”	#	$	%	&	’
4x	(	)	*	+	,	-	.	/	0	1
5x	2	3	4	5	6	7	8	9	:	;
6x	¡	=	¿	?	@	A	B	C	D	E
7x	F	G	H	I	J	K	L	M	N	O
8x	P	Q	R	S	T	U	V	W	X	Y
9x	Z	[	“	]	ˆ	˙	‘	a	b	c
10x	d	e	f	g	h	i	j	k	l	m
11x	n	o	p	q	r	s	t	u	v	w
12x	x	y	z	–	—	˝	˜	¨		

Bold, slanted, and sans-serif fonts follow the same layout. The italic fonts follow the same layout with an exception: italics contain the pound-sign character *£* instead of the dollar-sign character $. When using the italic font in Standard TEX, use `{\sl\$}` instead of \$.

Unlike Computer Modern Fonts, MicroPress Vector fonts all follow the same layout. Below is the layout of the `mvtmsr` font:

mvtmsr

	0	1	2	3	4	5	6	7	8	9
	˜	”	“	'	¡	¿	˝	˙	–	—
1x	ˆ	ff	fi	fl	ffi	ffl	ı	ȷ	`	´
2x	ˇ	˘	¯	˚	¸	ß	æ	œ	ø	Æ
3x	Œ	Ø		!	"	#	$	%	&	’
4x	(	)	*	+	,	-	.	/	0	1
5x	2	3	4	5	6	7	8	9	:	;
6x	<	=	>	?	@	A	B	C	D	E
7x	F	G	H	I	J	K	L	M	N	O
8x	P	Q	R	S	T	U	V	W	X	Y
9x	Z	[	\	]	^	_	‘	a	b	c
10x	d	e	f	g	h	i	j	k	l	m
11x	n	o	p	q	r	s	t	u	v	w
12x	x	y	z	{	|	}	~	¨	¶	§
13x	‹	›	«	»	†	‡	‚	„	£	¢
14x	*f*	‰	¥	Þ	þ	ð	Đ	Ł	đ	ł
15x	ij	IJ	©	®	™	•	●	⬤	■	□
16x	¤	☼	°	˛	…				▪	
17x	÷	±	·	μ	¬	ª	º	⁄	≤	≥

Notice that the layout includes all standard ASCII characters and punctuation Carrying characters such as ≤ or ¶ in every font may seem redundant; however it is necessary for many display fonts and even for the text fonts when one wants to distinguish between bold and light forms. Two display fonts below illustrate unusual forms of these characters:

mvfuts

	0	1	2	3	4	5	6	7	8	9
	ˉ	”	“	'	¡	¿	˝	˙	–	—
1x	ˆ	ff	fi	fl	ffi	ffl	ı	ȷ	`	´
2x	ˇ	˘	¯	˙	¸	ß	æ	œ	ø	Æ
3x	Œ	Ø		!	"	#	$	%	&	’
4x	(	)	*	+	,	-	.	/	0	1
5x	2	3	4	5	6	7	8	9	:	;
6x	<	=	>	?	@	A	B	C	D	E
7x	F	G	H	I	J	K	L	M	N	O
8x	P	Q	R	S	T	U	V	W	X	Y
9x	Z	[	\\	]	^	_	‘	a	b	c
10x	d	e	f	g	h	i	j	k	l	m
11x	n	o	p	q	r	s	t	u	v	w
12x	x	y	z	{	|	}	~	¨	¶	§
13x	‹	›	«	»	†	‡	‚	„	£	¢
14x	ƒ	‰	¥	Þ	þ	ð	Đ	Ł	đ	ł
15x	ĳ	Ĳ	©	®	™	•	●	●	■	□
16x	¤	☼	°	˛	…				■	
17x	÷	±	·	µ	¬	ª	º	⁄	≤	≥

mvcloi

	0	1	2	3	4	5	6	7	8	9
	˜	”	“	'	¡	¿	˝	˙	–	—
1x	ˆ	ff	fi	fl	ffi	ffl	ı	ȷ	`	´
2x	ˇ	˘	¯	˚	¸	ß	æ	œ	ø	Æ
3x	Œ	Ø		!	"	#	$	%	&	'
4x	(	)	*	+	,	-	.	/	0	1
5x	2	3	4	5	6	7	8	9	:	;
6x	<	=	>	?	@	A	B	C	D	E
7x	F	G	H	I	J	K	L	M	N	O
8x	P	Q	R	S	T	U	V	W	X	Y
9x	Z	[	\	]	^	_	'	a	b	c
10x	d	e	f	g	h	i	j	k	l	m
11x	n	o	p	q	r	s	t	u	v	w
12x	x	y	z	{	|	}	~	¨	¶	§
13x	‹	›	«	»	†	‡	‚	„	£	¢
14x	ƒ	‰	¥	Þ	þ	ð	Ð	Ł	đ	ł
15x	ĳ	Ĳ	©	®	™	•	●	●	■	□
16x	¤	☼	°	‘	…				■	
17x	÷	±	·	μ	¬			/	≤	≥

Uppercase Greek letters, when present, occupy positions 200 through 210.

11.3 Line Draw Font

The Line Draw fonts, mvttld and mvttldr, offer a quick and dirty way to typeset simple tables. These fonts mimic the ASCII line-draw characters. The characters in the first font have sharp corners, while in the second font they are rounded. The layout of these fonts is as follows:

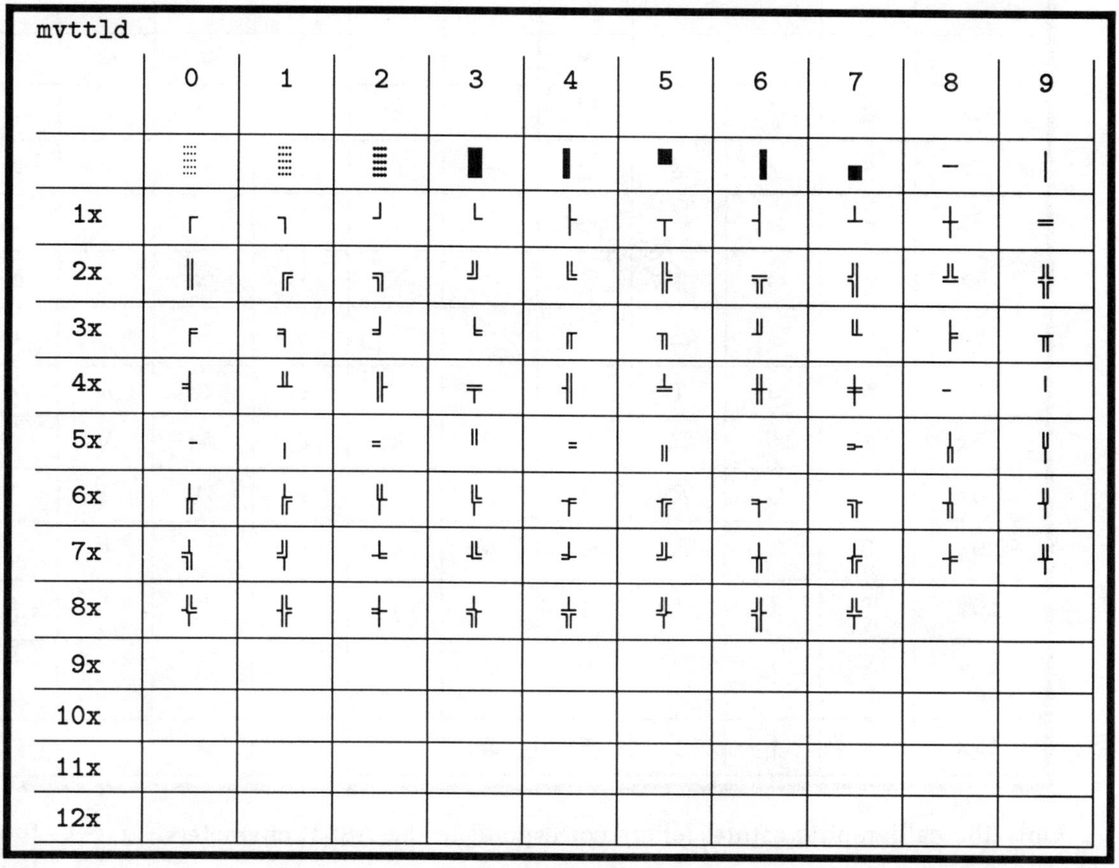

11.4 Symbol Font Layouts

The math-symbol font cmsy10 and extra-symbol fonts cmex10 contain symbols for use in formulas. These fonts contain 128 character each. Here is the layout of cmsy10:

cmsy10

	0	1	2	3	4	5	6	7	8	9
	−	·	×	∗	÷	⋄	±	∓	⊕	⊖
1x	⊗	⊘	⊙	○	∘	•	≍	≡	⊆	⊇
2x	≤	≥	⪯	⪰	∼	≈	⊂	⊃	≪	≫
3x	≺	≻	←	→	↑	↓	↔	↗	↘	≃
4x	⇐	⇒	⇑	⇓	⇔	↖	↙	∝	′	∞
5x	∈	∋	△	▽	/	ˈ	∀	∃	¬	∅
6x	ℜ	ℑ	⊤	⊥	ℵ	𝒜	ℬ	𝒞	𝒟	ℰ
7x	ℱ	𝒢	ℋ	ℐ	𝒥	𝒦	ℒ	ℳ	𝒩	𝒪
8x	𝒫	𝒬	ℛ	𝒮	𝒯	𝒰	𝒱	𝒲	𝒳	𝒴
9x	𝒵	∪	∩	⊎	∧	∨	⊢	⊣	⌊	⌋
10x	⌈	⌉	{	}	⟨	⟩	∣	‖	↕	⇕
11x	\	≀	√	∐	∇	∫	⊔	⊓	⊑	⊒
12x	§	†	‡	¶	♣	♢	♡	♠		

Only the calligraphic capital letters correspond to the ASCII characters.

The cmex10 font contains only large symbols (this font is shown reduced 40%):

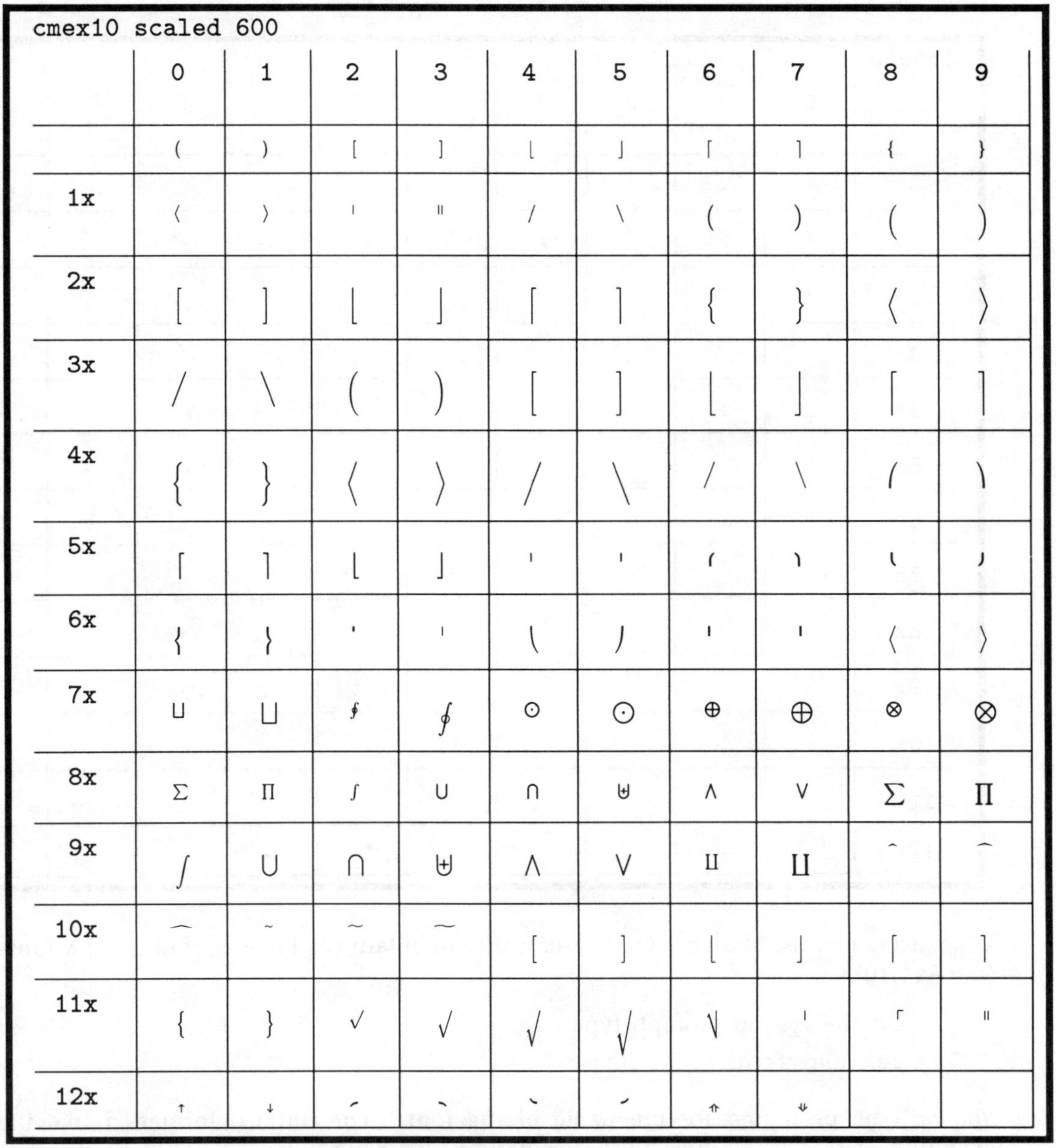

MicroPress Vector variants of these fonts (mvsy10 and mvex10) have identical layout.

Some symbol fonts may contain fewer than 128 characters. Here is the layout of a LaTeX font `line10`:

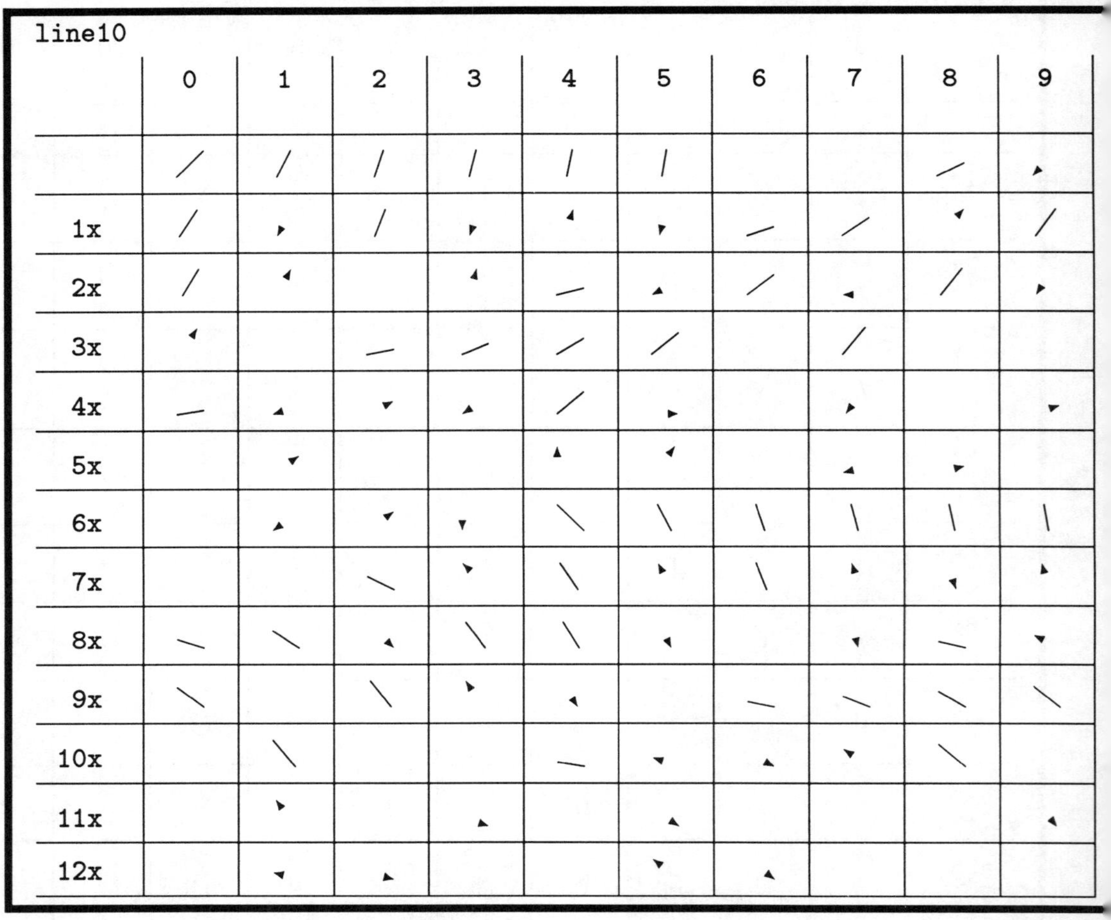

VTeX users can use the `showfont.tex` utility to obtain the layouts of other TeX fonts (see §11.10).

> On the system prompt, type
>
> `tex showfont`

You will be prompted for the name of the font. The output, formatted like the examples in this appendix, will be written to `showfont.dvi`.

11.5 MicroPress Vector Fonts – Partial Listing

mvaahb: A B C D E F G a b c d e f g 1 2 3 4 5 $ &

mvactb: A B C D E F G a b c d e f g 1 2 3 4 5 $ &

mvacti: A B C D E F G a b c d e f g 1 2 3 4 5 $ &

mvactj: A B C D E F G a b c d e f g 1 2 3 4 5 $ &

mvactr: A B C D E F G a b c d e f g 1 2 3 4 5 $ &

mvameb: A B C D E F G a b c d e f g 1 2 3 4 5 $ &

mvamei: A B C D E F G a b c d e f g 1 2 3 4 5 $ &

mvamej: A B C D E F G a b c d e f g 1 2 3 4 5 $ &

mvamer: A B C D E F G a b c d e f g 1 2 3 4 5 $ &

mvatcb: A B C D E F G a b c d e f g 1 2 3 4 5 $ &

mvatcr: A B C D E F G a b c d e f g 1 2 3 4 5 $ &

mvatpb: A B C D E F G a b c d e f g 1 2 3 4 $ &

mvatpr: A B C D E F G a b c d e f g 1 2 3 4 5 $ &

mvavnb: A B C D E F G a b c d e f g 1 2 3 4 5 $ &

mvavnd: A B C D E F G a b c d e f g 1 2 3 4 5 $ &

mvavnm: A B C D E F G a b c d e f g 1 2 3 4 5 $ &

mvavnr: A B C D E F G a b c d e f g 1 2 3 4 5 $ &

mvbdps: A B C D E F G a b c d e f g 1 2 3 $ &

mvblip: A B C D E F G a b c d e f g 1 2 3 4 5 $ &

mvbodb: A B C D E F G a b c d e f g 1 2 3 4 5 $ &

mvbodi: A B C D E F G a b c d e f g 1 2 3 4 5 $ &

mvbodj: A B C D E F G a b c d e f g 1 2 3 4 5 $ &

mvbodr: A B C D E F G a b c d e f g 1 2 3 4 5 $ &

mvbokd: A B C D E F G a b c d e f g 1 2 3 4 5 $ &

mvboke: A B C D E F G a b c d e f g 1 2 3 4 5 $ &

mvboki: A B C D E F G a b c d e f g 1 2 3 4 5 $ &

mvbokl: A B C D E F G a b c d e f g 1 2 3 4 5 $ &

mvbrod: A B C D E F G a b c d e f g 1 2 3 $ &

mvbrsh: A B C D E F G a b c d e f g 1 2 3 4 5 $ &

mvbskb: A B C D E F G a b c d e f g 1 2 3 4 5 $ &

mvbski: A B C D E F G a b c d e f g 1 2 3 4 5 $ &

mvbskj: A B C D E F G a b c d e f g 1 2 3 4 5 $ &

mvbskr: A B C D E F G a b c d e f g 1 2 3 4 5 $ &

mvchab: A B C D E F G a b c d e f g 1 2 3 4 5 $ &

mvchai: A B C D E F G a b c d e f g 1 2 3 4 5 $ &

mvchaj: A B C D E F G a b c d e f g 1 2 3 4 5 $ &

mvchar: A B C D E F G a b c d e f g 1 2 3 4 5 $ &

mvchcm: A B C D E F G a b c d e f g 1 2 3 4 5 $ &

mvclab: A B C D E F G a b c d e f g 1 2 3 4 $ &

mvclar: A B C D E F G a b c d e f g 1 2 3 4 5 $ &

mvcloi: A B C D E F G a b c d e f g 1 2 3 4 5 $ &

mvcorn: A B C D E F G a b c d e f g 1 2 3 4 5 $ &

mvcppb: A B C D E F A B C D E F 1 2 $ &

mvctnw: A B C D E F G A B C D E F G 1 2 3 4 5 $ &

mvdomc: A B C D E F G a b c d e f g 1 2 3 4 5 $ &

mveurb: A B C D E F G a b c d e f g 1 2 3 4 5 $ &

mveurr: A B C D E F G a b c d e f g 1 2 3 4 5 $ &

mvfgcr: A B C D E F G a b c d e f g 1 2 3 4 5 $ &

mvfgrr: A B C D E F G a b c d e f g 1 2 3 4 $ &

mvfgxr: A B C D E F G a b c d e f g 1 2 3 4 5 $ &

mvfscr: A B C D E F G a b c d e f g 1 2 3 4 5 $ &

mvfutb: A B C D E F G a b c d e f g 1 2 3 4 5 $ &

mvfutd: A B C D E F G a b c d e f g 1 2 3 4 5 $ &

mvfutl: A B C D E F G a b c d e f g 1 2 3 4 5 $ &

mvfutm: A B C D E F G a b c d e f g 1 2 3 4 5 $ &

mvfutr: A B C D E F G a b c d e f g 1 2 3 4 5 $ &

mvfuts: A B C D E F G a b c d e f g 1 2 3 4 5 $ &

mvfutx: A B C D E F G a b c d e f g 1 2 3 $ &

mvgacb: A B C D E F G a b c d e f g 1 2 3 4 5 $ &

mvgaci: A B C D E F G a b c d e f g 1 2 3 4 5 $ &

mvgacj: A B C D E F G a b c d e f g 1 2 3 4 5 $ &

mvgacr: A B C D E F G a b c d e f g 1 2 3 4 5 $ &

mvgalb: A B C D E F G a b c d e f g 1 2 3 4 5 $ &

mvgali: A B C D E F G a b c d e f g 1 2 3 4 5 $ &

mvgalj: A B C D E F G a b c d e f g 1 2 3 4 5 $ &

mvgalr: A B C D E F G a b c d e f g 1 2 3 4 5 $ &

mvgarb: A B C D E F G a b c d e f g 1 2 3 4 5 $ &

mvgari: A B C D E F G a b c d e f g 1 2 3 4 5 $ &

mvgarj: A B C D E F G a b c d e f g 1 2 3 4 5 $ &

mvgarr: A B C D E F G a b c d e f g 1 2 3 4 5 $ &

mvgdyb: A B C D E F G a b c d e f g 1 2 3 4 5 $ &

mvgdyr: A B C D E F G a b c d e f g 1 2 3 4 5 $ &

mvhand: A B C D E F G a b c d e f g 1 2 3 4 5 $ &

mvhlvb: A B C D E F G a b c d e f g 1 2 3 4 5 $ &

mvhlvl: A B C D E F G a b c d e f g 1 2 3 4 5 $ &

mvhlvr: A B C D E F G a b c d e f g 1 2 3 4 5 $ &

mvhlvx: A B C D E F G a b c d e f g 1 2 3 4 $ &

mvhobo: A B C D E F G a b c d e f g 1 2 3 4 5 $ &

mvjunr: A B C D E F G A B C D E F G 1 2 3 4 5 $ &

mvkorb: A B C D E F G a b c d e f g 1 2 3 4 5 $ &

mvkori: A B C D E F G a b c d e f g 1 2 3 4 5 $ &

mvkorj: A B C D E F G a b c d e f g 1 2 3 4 5 $ &

mvkorr: A B C D E F G a b c d e f g 1 2 3 4 5 $ &

mvlatw: A B C D E a b c d e 1 $ &

mvlthb: A B C D E F G A B C D E F G 1 2 3 4 $ &

mvltxt: A B C D E F G a b c d e f g 1 2 3 4 5 $ &

mvlubb: A B C D E F G a b c d e f g 1 2 3 4 5 $ &

mvlubd: A B C D E F G a b c d e f g 1 2 3 4 5 $ &

mvlubm: A B C D E F G a b c d e f g 1 2 3 4 5 $ &

mvlubr: A B C D E F G a b c d e f g 1 2 3 4 5 $ &

mvmelb: A B C D E F G a b c d e f g 1 2 3 4 5 $ &

mvmeli: A B C D E F G a b c d e f g 1 2 3 4 5 $ &

mvmelj: A B C D E F G a b c d e f g 1 2 3 4 5 $ &

mvmelr: A B C D E F G a b c d e f g 1 2 3 4 5 $ &

mvoptb: A B C D E F G a b c d e f g 1 2 3 4 5 $ &

mvoptr: A B C D E F G a b c d e f g 1 2 3 4 5 $ &

mvpalb: A B C D E F G a b c d e f g 1 2 3 4 5 $ &

mvpali: A B C D E F G a b c d e f g 1 2 3 4 5 $ &

mvpalj: A B C D E F G a b c d e f g 1 2 3 4 5 $ &

mvpalr: A B C D E F G a b c d e f g 1 2 3 4 5 $ &

mvpark: A B C D E F G a b c d e f g 1 2 3 4 5 $ &

mvpres: A B C D E F G a b c d e f g 1 2 3 4 5 $ &

mvrevu: A B C D E F G a b c d e f g 1 2 3 4 5 $ &

mvrokb: A B C D E F G a b c d e f g 1 2 3 4 5 $ &

mvrokm: A B C D E F G a b c d e f g 1 2 3 4 5 $ &

`mvrokx:` A B C D E F G a b c d e f g 1 2 3 4 $ &

`mvsvnb:` A B C D E F G a b c d e f g 1 2 3 4 5 $ &

`mvsvni:` A B C D E F G a b c d e f g 1 2 3 4 5 $ &

`mvsvnj:` A B C D E F G a b c d e f g 1 2 3 4 5 $ &

`mvsvnr:` A B C D E F G a b c d e f g 1 2 3 4 5 $ &

`mvtekr:` A B C D E F G a b c d e f g 1 2 3 4 5 $ &

`mvtifb:` A B C D E F G a b c d e f g 1 2 3 $ &

`mvtifi:` A B C D E F G a b c d e f g 1 2 3 4 5 $ &

`mvtifj:` A B C D E F G a b c d e f g 1 2 $ &

`mvtifr:` A B C D E F G a b c d e f g 1 2 3 4 5 $ &

`mvtmsb:` A B C D E F G a b c d e f g 1 2 3 4 5 $ &

`mvtmsi:` A B C D E F G a b c d e f g 1 2 3 4 5 $ &

`mvtmsj:` A B C D E F G a b c d e f g 1 2 3 4 5 $ &

`mvtmsr:` A B C D E F G a b c d e f g 1 2 3 4 5 $ &

`mvtroj:` A B C D E F G A B C D E F G 1 2 3 $ &

`mvunib:` A B C D E F G a b c d e f g 1 2 3 4 $ &

`mvunir:` A B C D E F G a b c d e f g 1 2 3 4 5 $ &

`mvurmr:` A B C D E F G a b c d e f g 1 2 3 4 5 $ &

`mvvagr:` A B C D E F G a b c d e f g 1 2 3 4 5 $ &

`mvwinr:` A B C D E F G a b c d e f g 1 2 3 4 $ &

All italic fonts are shown unslanted.

11.6 Defaults

The following is the list of some of TeX defaults:

Text Font ....roman, cmr10 (mvr10)

Math Font ....roman, cmmi10 (mvmi10)

Font Size ....10 pt

Magnification (\magnification) 1000

Page vertical size (\vsize) ....8.8 in

Page horizontal size (\hsize) ....6.5 in

Page vertical offset (\voffset) ....0 in

Page horizontal offset (\hoffset) ....0 in

Extra line-length allowance (\hfuzz) ....0.1 pt

Extra page-length allowance (\vfuzz) ....0.1 pt

Page number (\pageno) start with 1

Headline (\headline) ....empty

Footline (\footline) ....centered page number

Sentence spacing ....\nonfrenchspacing

Justification ....right

Paragraph indentation (\parindent) ....20 pt

Hanging indentation amount (\hangindent) ....0 pt

Hanging indentation lines (\hangafter) 1

Tolerance (\tolerance) 200

Horizontal badness (\hbadness) 1000

Line spacing (\baselineskip) 12pt

Small skip (\smallskip) ....3pt plus 1pt minus 1pt

Medium skip (\medskip) ....6pt plus 2pt minus 2pt

Big skip (\bigskip) ....12pt plus 4pt minus 4pt

11.7 Generating and Using Format Files

Most versions of TeX precompile macro and font declarations into *format files*. Since the format files can be simply loaded from disk without the need to re-read and re-process the same definitions all the time, they substantially improve the performance of the system. Format files can be created in the IniTeX mode of TeX. With most versions of TeX, VTeX included, IniTeX mode is invoked by specifying the `-i` switch on the command line. Some versions of TeX require a separate program.

Most versions of TeX include the source code of the standard macro package, `PLAIN.TEX`. This file contains the definitions of the majority of TeX commands described in this book; the exceptions are the commands that are actually built-in. Without the IniTeX facility, every document would need to include

```
input plain.tex
```

as its first line. Every time the document is typeset, TeX would need to re-read all the definitions included in `plain.tex`, some 5000 lines of source text. IniTeX allows to precompile `plain.tex` into the `plain.fmt` file, which later can be read fairly rapidly.

Full listing of the `plain.tex` file can be found in an Appendix to the TeXbook.

To precompile `plain.tex`, one enters

```
tex -i plain \dump
```

or, on some systems,

```
initex plain \dump
```

The `\dump` command first stores all the internal tables into the `plain.fmt` file. It then performs an `\end`.

VTeX implementation supplements `plain` with a number of additional macros. The `VTEX.TEX` file, used to create VTeX format, contains the following two lines:

```
\input plain   % Standard TeX macros
\input changes % VTeX enhancements
```

The `changes.tex` is fully described in the next section. Running IniTeX as shown above generates the `vtex.fmt` file.

Standard TeX implementations *automatically* load the `plain.fmt` file (VTeX loads `vtex.fmt`), unless they are run in the IniTeX mode or a different format file is specified. A different format file can be created and loaded with these steps:

1. Create a file containing the definitions you want to add to TeX. In most cases, you would want to start the file with the `\input plain` line. End the file

with \dump, not \end or \bye. For example, create the `eeffmt.tex` file containing

```
\input plain % or vtex
\def\EEF\{\bf Exotic and Edible Foods}
\def\EEFlogo{$\cal E\cal E\cal F$}
\def\boss{A.~Pumpernickel~3\raise.5ex\hbox{\sl rd}}
\dump
```

2. At the system prompt, type

```
tex -i eeffmt
```

This will create the `eeffmt.fmt` format file, that will contain all the traditional TeX commands as well as the new commands defined in `eeffmt.tex`.

➡ The \dump command is supported only in the InITeX mode.

You can use alternative formats, such as `eeffmt.fmt`, by specifying the format name preceded by an ampersand character `&` on the `tex` command line. For instance, you can type

```
tex &eeffmt reports
```

This will load the larger `eeffmt` format instead of the default `text` format. If the system cannot find the `eeffmt.fmt` file (it will search first the current directory and then the `text?` path), a warning message will be printed, and the `plain` (`vtex`) format will be loaded.

11.8 Graphics Inclusion in TeX and VTeX

TeX standard makes no a priori provisions for insertions of graphics files. There are good reasons for this: there is a multitude of possible graphics formats and no generally accepted standard. However, since a publishing system would be incomplete without some way to insert graphics, most TeX implementations provide at least a limited support. There are three distinct ways to handle graphics.

1. Support graphics files on the level of printing only. The user is responsible for allocating appropriate space within the document; the document further contains a `\special` command that is ignored by TeX itself but is handled by appropriate printer drivers. For example, if one wishes to insert a PostScript picture file, say, `picture.eps`, then the `.tex` file would contain something like

```
\vbox to 2in{\hbox to 3in{\special{ps:picture.eps}}
```

where the special command will be processed by the PostScript driver causing insertion of the picture at the current position. Other drivers are supposed to ignore this special. The dimensions of the surrounding box (2 and 3 inches) are usually computed manually. Virtually all PostScript printer drivers support this or similar syntax. Similarly, PCL (HP LaserJet series) pictures can be inserted with the `\special{hp:...}` syntax.

The disadvantage of this approach is first, printer dependence `.eps` pictures are printable only on PostScript, PCL only on LaserJet's, and neither can be previewed. Secondly, it is necessary to manually compute the space required for the picture. In the case of PostScript, the second problem can be greatly alleviated by the `PSFIG` collection of macros by Trevor Darrell.

2. An alternative approach is to convert the graphics files to bitmap fonts. Naturally, this could work only with bitmap graphics, and will require an extra conversion pass. Graphics files need to be maintained in two formats: the original (TIFF, PCX, PCL) and as a TEX-compatible font (PXL, PK). However, this approach takes care of the necessity of manually computing the amount of space needed: in the process of conversion of a graphics file into the TEX font format, the converter also creates a matching `.tfm` file. This approach is incompatible with VTEX, since the latter requires vector-font formats.

3. Finally, VTEX implements the `\makepicbox` graphics-insertion macro. `\makepicbox` takes three arguments: the horizontal and vertical resolutions of the picture and the name of the graphics file. For example, typing

   ```
   \makepicbox{300}{300}{whouse.mcp}
   ```

   tells TEX to retrieve a graphics file stored in the `whouse.mcp` file, treating it as a file at 300×300 dpi resolution. VTEX retrieves the actual dimensions from the graphics file itself and places the image into `\box0`. VTEX compares the original resolution with the printing resolution of the device and rescales the picture, if needed, to preserve its original size. Subsequent commands can position the image as desired. For example, to flush it right, one can now enter

   ```
   \rightline{\box0}
   ```

   to obtain

`\makepicbox` itself is a macro; it is explained in §11.9. Several bitmap formats are supported in this fashion, including MacPaint, PCX, and TIFF.

An additional common problem with graphics arises in handling the colors. To start with, many graphics pictures prepared with paint programs need to be inverted when printed. VTEX uses the `\special{I+}` and `\special{I-}` commands that turn the inversion of subsequent graphics files on and off. Furthermore, many graphics files are done in more than two colors. By default, VTEX treats any non-white color as black. Therefore, a blue-on-yellow picture is printed as solidly black. Since in most graphics formats different colors are stored as different color planes, VTEX allows you to select the desired planes. The `\special{C#}` command specifies which planes (primary colors) should be mapped to black and which ones are to be ignored. For example, `\special{C13}` forces VTEX to print color planes 1, 3 and 4 (here 13 is 1101 in binary). This special command is of particular use in color separation, since it allows printing the color planes separately. Color separation with TEX is intensively treated in Dan Levin's article in the *TUGboat,* magazine, Vol. 13, #2.

11.9 The changes.tex file

This appendix describes the contents of the `changes.tex` file which supplements the Plain format in the VTEX implementation. This listing is not necessarily complete; just enough is given to show the flavor. The actual listing is in teletype; the comments are in tempo.

```
% Lists

\def\list{\item{$\bullet$}}
\newcount \enumcount  \newcount \enumenumcount
\def\enum{\enumenumcount=96 \item{\the\enumcount.}
   \advance\enumcount by 1{}}
\def\enumenum{\advance\enumenumcount by 1
   \itemitem{\char\the\enumenumcount)}}

% Footnotes
\newcount\footcount
\def\nfootnote#1{\footnote{$^{\the\footcount}$}
  {#1}\advance\footcount by 1}

% Date
\def\today{\ifcase\month\or
  January\or February\or March\or April\or
   May\or June\or July\or August\or September\or
   October\or November\or December\fi
```

```
  \space\number\day, \number\year}
\def\upr#1{\uppercase\expandafter{#1}}
\def\page{\vfill\eject}

\newcount\a
\newcount\b
\def\dayofweek{
  \a=\year\advance\a by -1980\b=\a
    % Base computation of Jan 1, 1980 (Tuesday)
  \advance\b by 3
  \ifnum\month>2\advance\b by 1\fi
  \divide\b by 4              % Number of Leap Days passed
  \advance\a by \b            % Number of days passed.
  \advance\a by \ifcase\month
     \or1\or32\or60\or91\or121\or152%
     \or182\or213\or244\or274\or305\or335\fi
  \advance\a by \day
  \b=\a \divide\b by 7 \multiply\b by 7 \advance\a by-\b
  \ifcase\a Sunday\or Monday\or Tuesday\or
    Wednesday\or Thursday\or Friday \or Saturday\fi
  }
```

Notice that the above macro does not make corrections necessary to handle the years divisible by 100 and 400. Its working range covers years 1980 through 2099, when a new edition of this book may be needed.

```
% This is for use in legal agreements
\def\overbar#1{\setbox0=\hbox{#1}\vtop{\hbox{%
  \vrule height .4pt depth .4pt width \wd0} \par \box0}}

% Simple macros to correct alignments in \matrix, \cases
\def\rj#1{\hss#1\hfilneg}
\def\lj#1{\hfilneg#1\hss}

% Grey things
\def\graypattern#1{\def\grayset{M#1}}
\def\grayset{M01/00/10/00}
\def\gray{\special{\grayset}}                  %gray for boxes
\def\grayl{\special{M0001/0010/0100/1000}} %gray slanted left
\def\grayr{\special{M1000/0100/0010/0001}} %gray slanted right

% Fixed Obeyspaces
 \def\Obeyspaces{\catcode`\ =\active}
 {\Obeyspaces\global\let =\ }
```

....

The macros above need no explanation, since they have already occurred somewhere within the book. Those given below are more complicated and will receive special discussion. First, we will define the global attribute change commands, described in §2.8. Their implementation has a two-fold character: first, there are built-in integer variables `\fontscale`, `\outline`, `\shadow`, etc. Whenever a `\font` (or `\aliasfont`) command occurs, its parameters are *automatically* adjusted by taking these variables into consideration. For instance, given

```
\fontscale=1200 \fontslant=332 \outline=1
\font\test=cmr10 scaled 800 slant -100
```

`\test` will actually define `cmr10 scaled 960 slant 232 outline` (multiply scaling factors, add slants). If we now, for instance, define

```
\def\Large{\fontscale=1400}
```

Then `\Large` will work on all fonts declared afterwards, but will have no immediate effect. A better definition would be

```
\def\Large{\fontscale=1400\setfonts}
\def\setfonts{\the\font}
```

Notice that `\setfonts` re-issues the most recent font declaration. This definition will work just fine in the text mode; however it ignores the math fonts. `changes.tex` actually contains

```
\def\setmathfonts{%
  \textfont00\the\textfont00%
  \textfont01\the\textfont01%
  \textfont02\the\textfont02%
...
  \textfont15\the\textfont15%
  \scriptfont00\the\scriptfont00%
  \scriptfont01\the\scriptfont00%
...
  \scriptfont15\the\scriptfont15%
  \scriptscriptfont00\the\scriptscriptfont00%
  \scriptscriptfont01\the\scriptscriptfont01%
...
  \scriptscriptfont15\the\scriptscriptfont15%

 \newif\ifnewmath
 \def\setfonts{\the\font\newmathtrue}
 \everymath{\ifnewmath\setmathfonts\newmathfalse\fi}
 \everydisplay{\ifnewmath\setmathfonts\newmathfalse\fi}
```

and goes on to declare

```
\def\Huge{\fontscale=2400\setfonts}
\def\Large{\fontscale=1400\setfonts}
\def\BiG{\fontscale=1200\setfonts}
\def\Small{\fontscale=0800\setfonts}
\def\Tiny{\fontscale=0600\setfonts}
\def\Outline{\outline=1\setfonts}
\def\Slant#1{\slant=#1\setfonts}
```

and so on. It also makes possible the following footnote definition

```
\def\Footnote#1#2{\footnote#1{\Small #2}}
```

for printing footnotes at eight points. Note that

```
\def\Footnote#1{\footnote#1{\eightrm #2}}
```

is unacceptable since it would not allow fonts other than roman (and, therefore, math) in footnotes. The actual code in `changes.tex` is a bit more general: the definition of `\vfootnote` is changed to include

```
\fontscale=\footscale\setfonts
```

where

```
\newcount\footscale
\footscale=1000 % For compatibility
```

In a similar fashion one can (and `changes.tex` does) use `\Large` to re-define header fonts.

. . . .

We now present index-making macros, briefly described in §7.4 and §7.5. These were derived from the ones by Knuth, as appear in the *TEXbook*.

```
\catcode`@=11   %to be able to use private names
\newif\ifindextwice
```

The next five lines may be changed for customization. By default, we use 1-level index numbering with the `\pageno`. `\ientry` defines the format of an index entry. `\indexparams` may define additional options to be passed to the `IDXSRT` sorting program.

```
\indextwicefalse    %index is numbered by single numbers;
\font\@namefont=mvr10
\font\@reffont=mvtt10
\def\ientry#1#2{{\@namefont #1}\qquad {\@reffont #2}\par}
\def\iformat{\the\pageno}
\def\indexparams{}

\newdimen\pagewidth  \pagewidth=\hsize  % saved \hsize
```

```
\newdimen\pageheight \pageheight=\vsize  % saved \vsize
\newdimen\tmphsize

\def\pagesofar{\unvbox\partialpage
    \wd0=\hsize \wd2=\hsize
    \hbox to\pagewidth{\box0\hfil\box2}}

\newbox\partialpage

\def\begindoublecolumns{\begingroup
    \output={\global\setbox\partialpage=%
       \vbox{\unvbox255\bigskip}}\eject
    \output={\doublecolumnout}
    \tmphsize=\hsize \hsize=\pagewidth
    \setbox5=\vbox{\makeheadline} % initial value of headline
    \setbox6=\vbox{\makefootline} % initial value of footline
    \hsize=\tmphsize
    \hsize=\pagewidth  \advance\hsize by -1pc
      \divide\hsize by 2
    \vsize=\pageheight
    \advance\vsize by -\ht5
    \advance\vsize by -\ht6  % correct for footline / headline
    \multiply\vsize by 2    \advance\vsize by 1pc
     }
\def\enddoublecolumns{\output={\balancecolumns}\eject
    \endgroup
    \pagegoal=\pageheight}

\def\onepageout#1{\shipout\vbox{
    \offinterlineskip % butt the boxes together
    \tmphsize=\hsize \hsize=\pagewidth
      \setbox5=\vbox{\makeheadline}
      \setbox6=\vbox{\makefootline}
    \hsize=\tmphsize
      {\box5}
    \vbox to \pageheight{
      #1 % now insert the main information
      \ifvoid\footins\else % footnote info is present
        \vskip\skip\footins \kern-3pt
        \hrule height\ruleht width\pagewidth
           \kern-\ruleht \kern3pt
        \unvbox\footins\fi
      \boxmaxdepth=\maxdepth
      } % this completes the \vbox to \pageheight
      {\box6}
```

```
    }
  \advancepageno}

\def\doublecolumnout{\splittopskip=\topskip
    \splitmaxdepth=\maxdepth
  \dimen@=\pageheight \advance\dimen@ by-\ht\partialpage
  \setbox0=\vsplit255 to\dimen@ \setbox2=\vsplit255 to\dimen@
  \onepageout\pagesofar \unvbox255 \penalty\outputpenalty}

\def\balancecolumns{\setbox0=\vbox{\unvbox255} \dimen@=\ht0
      \advance\dimen@ by\topskip
      \advance\dimen@ by-\baselineskip
    \divide\dimen@ by2 \splittopskip=\topskip
    {\vbadness=10000 \loop \global\setbox3=\copy0
     \global\setbox1=\vsplit3 to\dimen@
     \ifdim\ht3>\dimen@ \global\advance\dimen@ by1pt \repeat}
    \setbox0=\vbox to \dimen@{\unvbox1}
    \setbox2=\vbox to\dimen@{\unvbox3}
    \pagesofar}
```

The \icopy and \iput place their argument into the index file. \iput removes it from the source; \icopy leaves it in. If the index file does not yet exist, it will be created.

```
\def\@indexname{}
\newwrite\@indexfile
\def\iput#1{%
  \expandafter\ifx\csname @indexopen \endcsname\relax%
    {%\message{opening file}%
    \immediate\write1{ }%
    \message{Index File [<cr> for \jobname.idx]: }%
    \immediate\read-1 to \@indexname%
       \def\ppar{\par}% if will be local & not last
       \ifx\ppar\@indexname\def\@indexname{\jobname.idx}\fi
    \immediate\openout\@indexfile=\@indexname
    \ifindextwice%
      \immediate\write\@indexfile{!INDEX=2}\else
      \immediate\write\@indexfile{!INDEX=1}\fi
    \global\def\csname @indexopen \endcsname{}%
       % we actually will not define, but make indexfile=fname
    \global\let\@indexname\@indexname
    }\fi%
  \write\@indexfile{#1 !0 \iformat.}%
  }
\def\icopy#1{#1\iput{#1}}    % this will copy and call put.
\catcode`@=12 % at signs are no longer letters
```

The user will use the \mergeindex command to actually include the index file in the document. \beginindex and \endindex turn on and off index-style formatting.

```
\def\beginindex{
  \catcode`@=11
  \begingroup
  \begindoublecolumns
  \parindent 0pt
  }

\def\endindex{\enddoublecolumns\endgroup\catcode`@=12}


\def\mergeindex{%
  \immediate\closeout\@indexfile%
  \command{idxsrt \indexparams\ \@indexname}%
  \beginindex
  \input index.tex      % Default output from IDXSRT
  \endindex
  \command{erase index.tex} %Clean up.
  }
```

....

We conclude the listing with a few macros responsible for the graphics insertion in VTEX. Notice that \sizegraph is a primitive command responsible for locating a picture file and retrieving its pixel dimensions into the \graphx and \graphy count variables.

```
 \newdimen\graphX
 \newdimen\graphY
 \def\scalegraph#1#2{
 \graphX=1in \divide\graphX by #1 \multiply\graphX by \graphx
 \graphY=1in \divide\graphY by #2 \multiply\graphY by \graphy
%\message{(Dots dimensions=\the\graphX;\the\graphY)}
                   }

\def\makepicbox#1#2#3{
\sizegraph #3
 %\message{(Pixel dimensions=\the\graphx;\the\graphy)}
  \ifnum\graphx=-1\errmessage{Cannot find picture file}\fi
  \ifnum\graphx=-2\errmessage{Invalid picture file}\fi
\scalegraph{#1}{#2}
\setbox0=\hbox{\special{G,#1,#2,#3}}
\ht0=\graphY
\wd0=\graphX
}
```

11.10 Printing Font Charts

This section contains the listing of the `showfont.tex` program used to prepare the font diagrams in this book. To use it, append

```
\def\fnt{...}\input showfont
```

in the body of your text. Replace the dots with the name of the font you want to print, together with the attributes. With some versions of TEX it may be needed to insert a `\page` immediately after `\input showfont`; failing to do so may cause them to run out of memory.

✔ When checking fonts for quality, it is recommended to run a series of tests at 4, 5, 6, 7, 8, 9, 11, 12, 14, and 16 points; the samples at 8 points and up should be printed with the `outline` option, since it makes glitches more apparent.

```
%% Uncomment the following line if you are not running VTeX:
%     \def\grayl{}
{
 % boxing declarations
\long\def\BOX#1{\parskip 0pt plus 1pt{
\hbox{     \grayp \vrule width 1mm
\vbox{     \grayp \hrule height 1mm
\hbox{ \kern5pt\vbox{\kern3pt#1\kern3pt}\kern5pt }
\grayp \hrule height 1mm%
}\grayp\vrule width 1mm
}}}
\long\def\BOXX#1%
  {{\hbadness10000\smallskip\nobreak\BOX{\vbox{#1}}}}
\catcode'\@=11
\font\F@nt\f0nt
 \def\lline#1#2{\tt#1x&&\gloop\char\charcount&&%
   \ifnum\charcount<#2%
   \global\advance\charcount1\repeat\cr
   \extra\noalign{\hrule}\extra
   \noalign{\global\advance\charcount1}
  }
\bigskip
\offinterlineskip
\def\extra{&height6pt&\omit&&\omit&&\omit
    &&\omit&&\omit&&\omit&&\omit
    &&\omit&&\omit&&\omit&&\omit\cr}
\def\hvi{\hskip6pt plus 1fill\relax}
\newcount\charcount\charcount\z@
\def\gloop#1\repeat{\gdef\body{#1}\iterate}
\BOXX{
```

```
\hbox{\tt \f0nt}\medskip
 \halign to\hsize{\hbox to40pt{\hfill#\hskip15pt}&
 \vrule#\tabskip\centering
 &\hvi\F@nt#\hvi
 &\vrule#&\hvi\F@nt#\hvi&\vrule#&\hvi
    \F@nt#\hvi&\vrule#&\hvi\F@nt#\hvi
 &\vrule#&\hvi\F@nt#\hvi&\vrule#&\hvi
    \F@nt#\hvi&\vrule#&\hvi\F@nt#\hvi
 &\vrule#&\hvi\F@nt#\hvi&\vrule#&\hvi
   \F@nt#\hvi&\vrule#&\hvi\F@nt#\hvi
 &\vrule#\tabskip0pt&#\cr
 \omit&height12pt&\gloop\tt\number
   \charcount&height12pt&\ifnum\charcount<9
  \global\advance\charcount1 \repeat\cr
 \extra\extra\extra\noalign{\hrule}
   \extra\noalign{\global\charcount\z@}
\lline{0}{9}                \lline{1}{19}
\lline{2}{29}               \lline{3}{39}
\lline{4}{49}               \lline{5}{59}
\lline{6}{69}               \lline{7}{79}
\lline{8}{89}               \lline{9}{99}
\lline{10}{109}             \lline{11}{119}
\lline{12}{129}             \lline{13}{139}
\lline{14}{149}             \lline{15}{159}
\lline{16}{169}             \lline{17}{179}
}} % BOXXx
}
```

11.11 Accompanying software

This book comes with a full-fledged demonstration version of the VTEX software in the IBM PC format. The demonstration diskette includes, in particular:

- VTEX typesetting program, VTEX.EXE
- previewer, VVIEW.EXE
- printer driver for the PostScript printers, VPOST.EXE
- printer driver for the LaserJet printers, VLASER.EXE
- printer driver for dot-matrix printers, VDOT.EXE
- intergrating shell, VT.EXE
- more than 30 typefaces

To use the software, you will need an IBM PC or compatible computer with 640 KB of RAM, 1.2MB 5.25" diskette drive and about 1.5MB of hard disk storage (3.50" version of the software is available on request from MicroPress, Inc.). To install the software, put the demonstration disk in the floppy drive and type

```
C> A:INSTALL
```

The install program will prompt you for the names of the directories to contain VTEX program and font files and copy the software to the hard disk. INSTALL will also pre-select the VGA monitor for the previewer and the LaserJet printer driver. Once the software has been installed, you should create the format file:

```
C> CD \VTEX
C> MAKEFMT
```

Run the VTEX shell

```
C> VT
```

to review or edit the printer/monitor selection. `VT` is a simple menu driven program; its main menu includes the following functions:

`Alt-N` Choose the Name of the document

`Alt-S` Set up the software

`Alt-E` Edit document. To use this feature, you should first select a text editor: choose Setup/Executables/Source Editor and enter the full name (with the path and the extension) of your editor program

`Alt-T` TEX the document

`Alt-P` Print

`Alt-V` Preview

`Alt-X` Exit to DOS

The commands can be entered with either keyboard or mouse. Help is available by pressing the `F1` key.

The following keys are used to navigate inside the previewer:

Keystroke	Action	Keystroke	Action
Esc	Quit VVIEW.	**I**	Inverse video.
+	Next page.	–	Previous page.
Ctrl ←	Left margin.	**Ctrl** →	Right margin.
Ins	Scroll left 10 pixels.	**Del**	Scroll right 10 pixels.
←	Scroll left 1 pixel.	→	Scroll right 1 pixel.
Home	Top of page.	**End**	Bottom of page.
PgUp	Screen up.	**PgDn**	Screen down.
↑	Scroll up 1 pixel.	↓	Scroll down 1 pixel.
S	Search text	**R**	Repeat search
C	Change Color	**B**	Change Background
1..9	Go to marker	**G**	Go to page

A collection of sample TEX files can be located in the `\VTEX\SAMPLES` subdirectory.

11.12 Comments, Please

Unlike TEX in general, VTEX is an evolving system. Users' suggestions over the last several years were extremely helpful for VTEX's development. The author of this book and VTEX's development team are very interested to hear readers' comments and suggestions about this book and the VTEX system. Please send your comments to

MicroPress, Inc.
68-30 Harrow Street
Forest Hills, NY 11375
Phone: 718-575-1816 Fax: 718-575-8038

Chapter 12

Index

\! **120**
¡ **45**
" **45**
\" **49**
\' **49**
” **45**
(< **45**
* **123**
\+ **221, 223**
\, **120**
- **45**
– **45**
— **45**
\. **49**
/’ **45**
/” **45**
\; **120**
\= **49**
\> **120**
>) **45**
>] **45**
¿ **45**
\aa **49**
\AA **49**
\above **100**
\abovewithdelims **100**
Active Character **178**
acute **49**
\acute **111**
\advance **137**
\advancepageno **60**
\ae **49**
\AE **49**
\afterassignment **185**
\aftergroup **185**
\aleph **248**
Alignment Tab **177**
\allowbreak **123**
\alpha **247**
alternative formats **269**
\amalg **249**
amoebae **50**
\angle **248**
APL **124**
\approx **250**
\arccos **104**
\arcsin **104**
\arctan **104**
\arg **104**
⟨argument⟩ **29**
\arrowvert **107**
\Arrowvert **107**
aspect **35, 38**
\Aspect **38**
\ast **249**
\asymp **250**
at **34**
\atop **100**
\atopwithdelims **100**
\b **49**
\backslash **107, 123, 124, 248**
badness **78**
\bar **111**
bar-under accent **49**
base line **84**
\baselineskip **84, 148**
Beginning of Group **177**
\beta **247**
\bf **23**
\big **107**
\Big **107**
big point **141**
\bigbreak **85**
\bigcap **249**
\bigcirc **249**
\bigcup **249**
\bigg **107**
\Bigg **107**
\biggl **107**
\Biggl **107**
\biggm **107**
\Biggm **107**
\biggr **107**
\Biggr **107**
\bigl **107**
\Bigl **107**
\bigm **107**
\Bigm **107**

\bigodot 249
\bigoplus 249
\bigotimes 249
\bigr 107
\Bigr 107
\bigskip 85, 150
\bigskipamount 85, 149, 150
\bigsqcup 249
\bigtriangledown 249
\bigtriangleup 249
\biguplus 249
\bigvee 249
\bigwedge 249
binary mod 104
Binary Operators 249
\bmod 104
bold 35
boldface 23
\boldsentence 161
\bot 248
\bowtie 250
\box 206
box variable 206
⟨box variable⟩ 206
bp 141
\bracevert 107
breve 49
\breve 111
\brokenrule 218
\brunderbar 218
\buildrel 113, 115
\bullet 249
\bye 7
\c 49
calligraphic capital letters 258
calligraphic letters 248
\cap 249
case-sensitive 13
\cases 116
\catcode 178, 179
category codes 176
cc 141
\cdot 249
\cdots 119
cedilla 49
\centerline 12, 59, 69
centimeter 141
\char 182
character categories 176
\chardef 182
\check 111
\chi 247
\choose 100
cicero 141
\circ 249
circumflex 49
\cleaders 212
\close 190
\clubsuit 248
cm 141
\colon 123, 124
\columns 221
\command 194
Comment 178
\cong 250
congruence 104
control sequences 7
\coprod 249
\copy 206
\cos 104
\cosh 104
\cosine 240
\cot 104
\coth 104
\count 137
⟨count⟩ 137
\cr 116, 221, 223, 224
Croatian 50
\csc 104
\csname 182
\cup 249
Czech 49
\d 49
\D 46
\dagger 249
dashes 45
\dashv 250
\day 137

dd **141**
\ddagger **249**
\ddot **111**
\ddots **119**
\def **155, 157, 159, 168**
default file names **7**
\deg **104**
Delimiters **251**
\delta **247**
\Delta **247**
depth **200**
⟨depth⟩ **200**
design size **34**
\det **104**
\determinant **167, 173**
\diamond **249**
\diamondsuit **248**
didot point **141**
dieresis **49**
\dim **104**
dimen **57**
\dimen **141**
⟨dimen⟩ **141**
\discretionary **76**
display math mode **95, 110, 199**
display style **111**
\displaystyle **110**
\div **249**
\divide **137**
dot **49**
\dot **111**
dot-under accent **49**
\doteq **250**
\dotfill **212**
\dots **45**
double superscript **97**
double-spaced **84**
\downarrow **107, 251**
\Downarrow **107, 251**
\downbracefill **212**
drop-case **210**
\dump **268**
\edef **170**
\eject **87**
\ell **248**
\else **171, 173**
em **141**
em-dash **143**
emergency stop **9**
\emptyset **248**
\end **7**
End of Group **177**
End of Line **177**
\endcsname **182**
\endinput **16**
\enspace **65, 161**
\enum **72**
\enumenum **72**
\epsilon **247**
\eqalign **122**
\eqno **121**
\equiv **250**
\errhelp **184, 189**
\errmessage **189**
\errno **194**
es-zet **50**
Escape **177**
\eta **247**
\everycr **184**
\everydisplay **184**
\everyhbox **184**
\everyjob **184**
\everymath **184**
\everypar **184**
\everyvbox **184**
ex **141**
\exec **194**
\exists **248**
\exp **104**
\expandafter **170**
ff **49**
ffi **49**
ffl **49**
fi **49**
\fi **171, 173**
fil **145, 149**
\filbreak **87**
fill **145, 149**

fillpattern 38
\Fillpattern 38
fl 49
\flat 248
\folio 60
\font 30
\fontscale 38
\fontsentence 161
\footcount 88, 137
\footins 88
\footline 60, 171, 184
\footnote 88
\forall 248
format file 268
fractions, complex 102
French 50
\frenchspacing 63
\frown 250
Fuehrer 50
\gamma 247
\Gamma 247
\gcd 104
\ge 250
\geq 250
German 50
\gets 251
\gg 250
\global 137, 140, 167, 168
⟨glue⟩ 150
\goodbreak 87
grave 49
\grave 111
\gray 215
\grayl 215
\graypattern 215
\grayr 215
\H 49
hacek 49
\halign 224
handwritten symbols 124
\hangafter 69, 210
\hangindent 69, 210
hanging indentation 69
hard space 75
\hat 111
\hbadness 78
\hbar 248
\hbox 75, 98, 202
\headline 60, 171, 184
\heartsuit 248
height 200
⟨height⟩ 200
hex 137
\hfil 145
\hfill 145
\hfilneg 145
\hfuzz 78
\hoffset 57
\hom 104
\hookleftarrow 251
\hookrightarrow 251
horizontal mode 84, 199
\hrule 28, 200
\hrulefill 212
\hsize 57, 204
\hskip 65
\hss 145
Hungarian 49
Hungarian umlaut 49
\hyphenation 76
\i 49
\if 182
\ifcase 173
\ifcat 182
\ifdim 171
\ifeof 171, 173
\iffalse 171, 172
\ifhbox 171
\ifhmode 171, 173, 199
\ifinner 171, 173, 199
\ifmmode 171, 173, 199
\ifnum 171, 172
\ifodd 171
\iftrue 171, 172
\ifvbox 171
\ifvmode 171, 173, 199
\ifvoid 171
\ifx 171

Ignored **177**
\Im **248**
\imath **248**
\immediate **190**
in **141**
\in **104, 250**
inch **141**
\indent **65, 67**
\inf **104**
\infty **248**
input **173**
\input **16**
\int **99**
internal vertical mode **199**
\intop **249**
Invalid **178**
invisible rule **202**
\iota **247**
\it **23**
Italian **49**
italic correction **26**
\item **72**
\itemitem **72**
\j **49**
\jmath **248**
\joinrel **113**
\kappa **247**
\ker **104**
\kern **208**
\l **49**
\L **49**
\lambda **247**
\Lambda **247**
\land **249**
landscape **58**
\landscape **57**
\langle **107, 251**
Large Math Operators **248**
Latin **50**
\lbrace **107, 251**
\lbrack **107, 251**
\lccode **180**
\lceil **107, 251**
\ldots **119**
\le **250**
\leaders **212**
\left **107**
\leftarrow **251**
\Leftarrow **251**
\leftarrowfill **212**
\leftharpoondown **251**
\leftharpoonup **251**
\lefthyphenmin **76**
\leftline **59, 69**
\leftrightarrow **251**
\Leftrightarrow **251**
\leftskip **69**
\leq **250**
\leqno **121**
\let **185**
Letter **178**
letterhead **150**
\lfloor **107, 251**
\lg **104**
\lgroup **107**
\lim **104**
\liminf **104**
\limits **99, 249**
\limsup **104**
\line **59, 69, 144, 204**
Line Draw **256**
\list **72**
\lj **119, 125**
\ll **250**
\lmoustache **107**
\log **104**
\long **168**
\longleftarrow **251**
\Longleftarrow **251**
\longleftrightarrow **251**
\Longleftrightarrow **251**
\longmapsto **251**
\longrightarrow **251**
\Longrightarrow **251**
\loop **175**
\lor **249**
\lower **208**
\lowercase **180**

m-width **141**
Macro Parameter **177**
macron **49**
\magnification **34**
\magstep **34**
\magstephalf **34**
\mapsto **251**
math examples **128**
math mode **199**
Math Shift **177**
\mathbin **123**
\mathchardef **126, 128**
\mathclose **123**
\mathinner **123**
\mathop **123**
\mathopen **123**
\mathord **123**
\mathpunct **123**
\mathrel **123**
\mathstrut **113, 114**
\matrix **116**
\max **104**
\medbreak **85**
medium space **120**
\medskip **85, 150**
\medskipamount **85, 149, 150**
\message **189**
\mid **250**
millimeter **141**
\min **104**
minus **143**
missing brace **168**
\mit **248**
mm **141**
mod **104**
\models **250**
\month **137**
\moveleft **208**
\moveright **208**
\mp **249**
mu **141**
mu **141**
\mu **247**
\multiply **137**
\multispan **228**
\nabla **248**
\narrower **69**
\natural **248**
\ne **250**
\nearrow **251**
\neg **248**
negative space **120**
\neq **250**
\newcount **137**
\newdimen **141**
\newglue **149**
\newif **175**
\newread **190**
\newskip **149**
newspaper-style **79**
\newtoks **182**
\newwrite **190**
\nfootnote **88**
\ni **250**
\nobreak **87**
\noexpand **170**
\noindent **67**
\nolimits **99, 249**
\nonfrenchspacing **63**
\nopagenumbers **60**
\not **250**
\notin **250**
\nu **247**
\null **63**
\number **137**
⟨number⟩ **137**
\nwarrow **251**
\o **49**
\O **49**
\obeylines **90, 178**
\obeyspaces **90**
\odot **249**
\oe **49**
\OE **49**
Oedipus Rex **50**
\offinterlineskip **228**
ogonek accent **49**
\ointop **249**

Old English **50**
old-style digits **248**
\omega **247**
\Omega **247**
\ominus **249**
\omit **228**
\openin **190**
\openout **190**
\oplus **249**
\or **173**
ordinary math mode **95, 110, 199**
\oslash **249**
Other **178**
\otimes **249**
out of memory **168**
\outer **168**
outline **38**
\Outline **38**
\output **184, 211**
\over **100**
\overbrace **113, 215**
\overleftarrow **113**
\overline **113**
\overrightarrow **113**
\overwithdelims **100**
\page **5, 87, 149**
\pageno **60, 137**
\par **67**
\parallel **250**
parenthesized mod **104**
\parindent **67, 72**
\parshape **69**
\parskip **84, 148**
\partial **248**
pattern specification **216**
pc **141**
\perp **250**
\phantom **124**
\phi **247**
\Phi **247**
\pi **247**
\Pi **247**
pica **141**
plus **143**
\pm **249**
\pmatrix **116, 147**
\pmod **104, 105**
point **141**
Polish **50**
Portuguese **49**
\pr **104**
\prec **250**
\preceq **250**
printable page **57**
\proclaim **160**
\prod **249**
\propto **250**
\psi **247**
\Psi **247**
pt **141**
\qquad **65, 120**
\quad **65, 120, 143**
quotes **45**
\raggedright **67**
\raise **208**
\random **174**
\rangle **107, 251**
\rbrace **107, 251**
\rbrack **107, 251**
\rceil **107, 251**
\Re **248**
\read **190**
redline **217**
Relations **250**
\relax **7**
repeat **175**
restricted horizontal mode **99, 199**
\rfloor **107, 251**
\rgroup **107**
\rho **247**
\right **107**
\rightarrow **251**
\Rightarrow **251**
\rightarrowfill **212**
\rightharpoondown **251**
\rightharpoonup **251**

\righthyphenmin 76
\rightleftharpoons 251
\rightline 59, 69
\rightskip 69
\rj 119, 125
\rm 23
\rmoustache 107
roman font 24
Romanian 49
\romannumeral 137
\root 105
rules 200
\sc 31
scaled 34
scaled point 141
scaled points 143
Scandinavian languages 50
script style 111
\scriptfont 126
scripts 96
scriptscript style 111
\scriptscriptfont 126
\scriptscriptstyle 110
\scriptstyle 110
\searrow 251
\sec 104
\setbox 206
\setminus 123, 124, 249
\settabs 221, 223
\sf 31
shadow 38
\Shadow 38
\sharp 248
sharp S 50
\shipout 211
\show 155, 178, 182
showfont.tex 260
\showhyphens 76
\showthe 137
shrink 143
\sigma 247
\Sigma 247
\sim 250
\simeq 250
\sin 104
\sincos 240
\sine 240
single-spaced 84
\sinh 104
\skew 111
\sl 23
slant 38
\Slant 38
slanted font 24
\slash 75
Slovak 49
\smallbreak 85
smallcaps 38
\Smallcaps 38
\smallskip 85, 150
\smallskipamount 85, 149, 150
\smile 250
sp 141
Space 177
\spadesuit 248
Spanish 46
\special 195
special characters 12
spread 204, 224
\sqcap 249
\sqcup 249
\sqrt 105
\sqsubseteq 250
\sqsupseteq 250
\ss 49
\star 249
stretch 143
strikeout 217
\string 182
\strut 200, 228
stylesheet 157
Subscript 177
\subset 250
\subseteq 250
\succ 250
\succeq 250
\sum 99, 249
\sup 104

Superscript **177**
\supset **250**
\supseteq **250**
\surd **248**
\swarrow **251**
Swedish **49**
\t **49**
\tabalign **221, 223**
\tabskip **224**
\tan **104**
\tanh **104**
\tau **247**
\TeX **12**
text italics font **24**
text macros **156**
text style **111**
\textfont **126**
\textindent **69**
\textstyle **110**
\the **137**
\theta **247**
\Theta **247**
thick space **120**
thin space **120**
tie-after accent **49**
tie-in **46, 75**
tilde **49**
\tilde **111**
\times **249**
to **204, 224**
\to **251**
\tolerance **78**
\top **248**
\triangle **248**
\triangleleft **249**
\triangleright **249**
true **38**
true **141, 142**
\tt **23**
typewriter font **24**
\u **49**
\uccode **180**
umlaut **49**
unbreakable space **75**
\underbar **28**
\underbrace **113, 215**
\underline **113**
\uparrow **107, 251**
\Uparrow **107, 251**
\upbracefill **212**
\updownarrow **107, 251**
\Updownarrow **107, 251**
\upleftarrow **215**
\uplus **249**
\uppercase **180**
\upr **137**
\uprightarrow **215**
\upsilon **247**
\Upsilon **247**
\v **49**
\varepsilon **247**
variable, boolean **175**
\varpi **247**
\varrho **247**
\varsigma **247**
\vartheta **247**
\vbadness **78**
\vbox **202, 205**
\vcenter **202, 206**
\vdash **250**
\vdots **119**
\vec **111**
\vee **249**
\vert **107**
\Vert **107**
vertical mode **84, 199**
\vfil **148**
\vfill **148**
\vfilneg **148**
\vfootnote **88**
\vfuzz **78**
visible space command **13**
\voffset **57**
\vrule **200**
\vsize **57**
\vskip **85**
\vss **148**
\vtop **202, 205**

\wedge **249**
\widehat **111**
\widetilde **111**
width **200**
⟨width⟩ **200**
\wp **248**
\wr **249**
\write **190**
x-height **141**
\xdef **170**
\xi **247**
\Xi **247**
\xleaders **212**
\year **137**
\zeta **247**
[< **45**
\‘ **49**
“ **45**